1 MONTH OF
FREE
READING

at

www.ForgottenBooks.com

By purchasing this book you are eligible for one month membership to ForgottenBooks.com, giving you unlimited access to our entire collection of over 1,000,000 titles via our web site and mobile apps.

To claim your free month visit:

www.forgottenbooks.com/free902995

ISBN 978-0-266-87590-1
PIBN 10902995

For support please visit www.forgottenbooks.com

PRACTICAL AND THEORETICAL SYSTEM

OF

ARITHMETIC,

CONTAINING

SEVERAL NEW METHODS OF OPERATION,

AND A

NEW SYSTEM OF PROPORTION;

WITH

THEORETICAL EXPLANATIONS OF ALL THE PRINCIPAL RULES.

ALSO,

A TREATISE ON MENSURATION,

AND A

BRIEF PRACTICAL SYSTEM

OF

BOOK-KEEPING.

BY GEORGE WILLSON.

CANANDAIGUA:

PUBLISHED AND SOLD, WHOLESALE AND RETAIL, BY

C. MORSE.

OLD ALSO BY KEESE, COLLINS & CO., N. & J. WHITE, AND LEAVITT,
LORD & CO., NEW YORK; OLIVER STEELE, AND W. C. LITTLE,
ALBANY; AND BENNET & BRIGHT, UTICA

1836

N 2 a -

PREFACE.

In constructing an Arithmetic for the use of schools, two errors are to be avoided; the total *exclusion* of explanation, and the *redundancy* of it :—the laying down of naked arbitrary rules, as if the scholar were incapable of comprehending the *reason* of them; and that minuteness and excess of explanation, which leaves nothing for his own discovery and investigation.

Explanations should, in the judgment of the Author, be applied to the *principle* of the rules, rather than to the solution of particular questions; and nothing should be done *for* the learner, which he is capable of doing for himself. Every one recollects the satisfaction, with which, as he advanced in his mathematical course, he found himself able to take the progressive steps by himself, without the aid of his teacher or his fellow-students; and every one is conscious of the surer grasp, he yet retains, on the acquisitions he made by the un-assisted exercise of his own faculties. To the mind as to the limbs, *leading strings* may be of service in tak-ing the first steps; but, if never laid aside, there can never be confidence in the ability to *go alone.*

In the preparation of this treatise, the Author has aimed especially at brevity, clearness and precision; and he has added only so much of theoretical explanation, as will enable the scholar to understand the *principle* and *reason* of the several rules, and methods of operation.

A new method of statement in Proportion, and several new and convenient contractions, under different rules, will be found in this book.

The reader is referred to Decimals, to the theoretical explanations of Proportion, the Square and Cube Roots, and to Mensuration, as containing original matter, not to be met with in other systems.

A brief practical system of Book-keeping is subjoined, which will be found sufficient for all the purposes of ordinary business; and indeed, for mercantile establishments, those only excepted, whose extensive operations require the use of the more complicated system of *double entry*.

As the best preparation for the use of this book, or of any other treatise of the same grade, the Author would recommend a thorough acquaintance with Col- burn's Mental Arithmetic. The effect of this invaluable little work, in disciplining the minds of younger pupils to habits of clear and accurate thinking, and the aptitude and rapidity it gives in numerical calculations, are known to all who have used the book.

A word in conclusion, on the method of teaching Arithmetic. No school should be without a *blackboard*. The scholars should be formed into classes, every member of which, should in rotation, be required to work a question on the black-board, in the presence of his class, and *to explain each step of the process;* and the reason of it. If taken through the book in this way, it is hardly possible, that a pupil should fail of attaining a competent knowledge of the science.

ARITHMETIC.

ARITHMETIC treats of the properties, relations, and combination of numbers.

Its principal branches are, Notation or Numeration, Addition, Subtraction, Multiplication, and Division.

Numeration teaches to write and read numbers.

METHOD OF NOTATION.

The characters by which numbers are expressed, were derived from Arabia. They are the following: 1, 2, 3, 4, 5, 6, 7, 8, 9, 0. By variously combining these, all possible numbers are expressed. The simple characters, however, carry us no higher in numbering than nine units. To denote an additional unit, the first figure is repeated, but in a higher place, having the cipher (which by itself has no value) on the right: the 1 then becomes a unit of the *second order*, and has ten times its simple value. The addition is continued up to nineteen, by placing the simple figures at the right of this unit of the second order, in the room of the cipher: thus, 11, 12, 13, 14, 15, 16, 17, 18, 19; which may be read, one ten and one, one ten and two, &c. When all the simple characters are repeated in this combination *once*, the addition is made by increasing the unit of the second order one, (making it two tens,) and repeating the cipher: thus, 20. Again the simple characters are added successively in the same manner as before, and we have 21, 22, 23, &c. When the whole are repeated, the second order is again increased, and becomes 30. After the nine digits have all in succession been used in the second order, we are carried in numbering up to 99, or, nine tens and nine; then the additional number is expressed, by removing the original unit one place farther to the left, where it becomes a unit of the third order, and has ten times the value which it had in the second, and one hundred times its simple value: it is then named hundreds

5

and all the preceding combinations are repeated successively with it, up to 199, when this unit of the third order is increased, and becomes 200.

The simple figures being used in succession, as units of the *third* order, bring us to 999, when another remove toward the left, gives them a tenfold higher value, and makes them *thousands*, (1000.) All the preceding combinations are then repeated, with their own appropriate names, and that, of thousand superadded : the notation thus carries us to six figures, ending with 999.000—(that is, nine hundred and ninety-nine thousand.) Then succeed three more places of figures, read like the first three, with the name of *million* superadded. Thus the name changes with every additional three places toward the left.

Three places of figures commencing at the right form a period, to which a distinctive name is given, as follows : Units, Thousands, Millions, Billions, Trillions, Quadrillions, Quintillions, Sextillions, Septillions, Octillions, Nonillions, Decillions, Undecillions, Duodecillions.

Every period is read precisely alike, except that a different name is added at the end of each.

I. *To read numbers,* therefore :

RULE. Beginning at the right, divide them into periods of three figures, and then read from the left each period by itself, adding to each its appropriate name.

EXAMPLE.

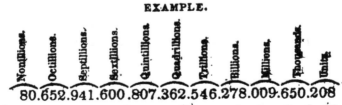

80.652.941.600.807.362.546.278.009.650.208

Which is read, eighty nonillions, six hundred fifty-two octillions, and so on to the last period, to which the name (units) is not added.

Divide off by points, and read the following numbers :

5461
412316
4666240
987000000
600588214927008360094

II. *To write numbers.*

RULE. Beginning at the left hand, write by periods, placing each period in its proper order; taking care to supply by ciphers, those periods and places, that are omitted in the question.

If, for example, the number to be written, be *ninety billions, four hundred and sixty-one thousand and twelve*— you begin at the left hand, and write the billions (90.) Next to billions comes the period of millions; but as there are no millions named in the question, you fill up that period with ciphers. Thousands follow millions, which you write (461), and units close the series. But as there are only two places of units in twelve, you fill out that period by a cipher in the place of hundreds, (012.)

The whole number stands thus: 90.000.461.012.

Write the following numbers:

Nine millions, seventy-two thousand, and two hundred.

Eight hundred millions, forty-four thousand, and fifty-five.

Eight billions, sixty-five millions, three hundred and four thousand, and seven.

Fifty-four sextillions, three hundred trillions, sixty-seven millions, four hundred and twenty.

Seventy decillions, two hundred and thirty-one octillions one billion, one hundred thousand, and three hundred.

From the system of notation already explained, it is evident that figures have a simple value and a local value.

When a figure stands in the place of units, it has a simple value.

When a figure does not stand in the place of units, it has a local value, which varies according to its distance from the unit's place.

In the annexed series, the first right hand figure has a simple value, the second has a local value tenfold greater: the third a still higher local value, tenfold greater than the second; and so on, increasing in the same proportion with every additional remove toward the left.

Millions	C Thousands	X Thousands	Thousands	Hundreds	Tens	Units
1	1	1	1	1	1	1
2	2	2	2	2	2	2

QUESTIONS.*

From whence are the numerical characters derived ?
What is the method of combining them to express numbers ?
What advantage do you perceive in the decimal system of notation ?
What do you suppose suggested the number 10, as the basis of numerical combination ?

---

* The design of the questions in this book, is to elicit thought and investigation on the part of the learner; not to save the teacher the labor of framing questions on the different rules and definitions. One of the principal objects of education, it is conceived, ought to be, to put the learner upon the exercise of his own faculties; an object but very little promoted, by the ordinary plan of constructing questions for school-books, in a way to direct the pupil almost *mechanically* to the answer to be given.

SIMPLE ADDITION.

ADDITION is putting together several smaller numbers to find their amount, or sum total : thus, 4 dollars and 8 dollars in one sum, are 12 dollars.

Numbers to be added to each other are sometimes connected by a sign, thus $6+4=10$: which is read, 6 plus (that is *more*) 4, are equal to 10. Two parallel lines are the sign of equality.

ADDITION TABLE.

$2+1=3$	$3+1=4$	$4+1=5$	$5+1=6$
$2+2=4$	$3+2=5$	$4+2=6$	$5+2=7$
$2+3=5$	$3+3=6$	$4+3=7$	$5+3=8$
$2+4=6$	$3+4=7$	$4+4=8$	$5+4=9$
$2+5=7$	$3+5=8$	$4+5=9$	$5+5=10$
$2+6=8$	$3+6=9$	$4+6=10$	$5+6=11$
$2+7=9$	$3+7=10$	$4+7=11$	$5+7=12$
$2+8=10$	$3+8=11$	$4+8=12$	$5+8=13$
$2+9=11$	$3+9=12$	$4+9=13$	$5+9=14$
$6+1=7$	$7+1=8$	$8+1=9$	$9+1=10$
$6+2=8$	$7+2=9$	$8+2=10$	$9+2=11$
$6+3=9$	$7+3=10$	$8+3=11$	$9+3=12$
$6+4=10$	$7+4=11$	$8+4=12$	$9+4=13$
$6+5=11$	$7+5=12$	$8+5=13$	$9+5=14$
$6+6=12$	$7+6=13$	$8+6=14$	$9+6=15$
$6+7=13$	$7+7=11$	$8+7=15$	$9+7=16$
$6+8=14$	$7+8=15$	$8+8=16$	$9+8=17$
$6+9=15$	$7+9=16$	$8+9=17$	$9+9=18$

RULE. Having placed units under units, tens under tens, &c., begin with the right hand column, and add it up. See how many tens are contained in the sum; set down what is over, and carry as many to the next column, as there were tens in the first: proceed in the same manner with the remaining columns, and at the last, set down the whole amount.*

(1)	(2)	(3)	(4)
53	291	6432	346977
52	851	9478	413339
13	152	1666	328012
89	698	7412	877543
Sum 207	1992	24988	1965871

The reason of carrying for the tens, is obvious from the system of notation already explained. Ten in an inferior column, is equal to but one in the next higher; and as you can write only the excess of the tens under any one column, the tens themselves must be carried to the next higher.

(5)	(6)
6 1 2 4 5 0 7 9 6 8 2	8 7 0 3 2 6 3 4 7 2 0 1 3
8 0 4 2 9 3 1 5 5 3 8	5 6 5 2 1 7 4 6 3 0 1 2 8
5 3 0 3 4 7 9 2 8 4 1	6 0 8 1 2 7 5 3 0 6 2 1 7
2 4 7 1 3 5 2 0	4 3 6 7 5 8 3 0 2 1 4 6 3
3 2 0 8 1 2	7 1 5 2 3 4 9 7 1 3 6 2 0

7. Add together 3426+9024+5106+8390+1204.
   Ans. 27150.
8. Add 509267+235809+72910+8392+420+21+9.
   Ans. 826828.
9. Add 19+817+4298+50916+730205+9120634.
   Ans. 9906889.
10. Add together the days of the twelve calendar months.
   Ans. 365

* To prove addition add the figures downwards.

11. An estate being shared equally between six heirs, the portion of each heir was 37542 dollars; what is the amount of the whole estate?   Ans. $ 225252.

12. An army consisting of 52714 infantry, 5110 cavalry, 6250 dragoons, 3927 light-horse, 928 gunners, 1410 pioneers, 250 sappers, and 406 miners; what is the whole number of the army?   Ans. 70995.

FEDERAL MONEY,

Or, the Currency of the United States.

TABLE.

10 mills (m) make	1 cent, marked	c.
10 cents	1 dime,	d.
10 dimes	1 dollar,	dol. or $.
10 dollars	1 Eagle,	E.

As the denominations of Federal Money increase in a tenfold proportion, they are added and subtracted like whole numbers. A point called a separatrix or decimal point, is placed at the right of the dollars, to separate them from the inferior denominations. In practice, dimes and eagles are disregarded in naming numbers. The dimes are read as cents, and the eagles as dollars : thus, $12.75 is, 12 dollars 75 cents ; not 1 eagle, 2 dollars, 7 dimes, and 5 cents,

RULE. Place dollars under dollars, cents under cents, &c. and add as in whole numbers ; and put the separatrix is the sum total, under the separating points above.

If the cents are less than 10, a cipher must be put in the place of dimes.

(1)	(2)	(3)
$ d c m	$ d c	d c m
139. 8 6 7	4319. 8 9	.9 7 6
1273. 5 9 4	3287. 8 0	.4 5 8
9807. 6 0	1829. 1 6	.6 2 9
7695. 2 4	130. 0 1	.5 8 3

$18916. 3 0 1

4. What is the sum of 140 dollars 9 cents, 24 dollars 16 cents, 5 dollars 25 cents, 5 dollars 5 cents, and 304 dollars 8 cents 9 mills ?   Ans. $478.639.

5. A man left his estate to his two children ; who, after payment of the debts amounting to 645 dollars 75 cents, received each 3842 dollars : . how much was the whole estate ?   Ans. $ 8329.75.

6. If I owe to A, 60 dollars and 8 cents ; to B, 12 dollars more than that sum ; and to C, 75 cents more than to both the others : how much do I owe them all ?

Ans. $ 265 07.

7. What is the amount of a bill, the items of which are, 2¼ yards of broadcloth at 8 dollars per yard ; 3 yards of kerseymere, at 10 dollars ; a vest pattern, at 2 dollars 50 cents, and trimmings for vest, 75 cents ?   Ans. $ 31.25

8. Three men were to pay a certain sum of money : A paid 25 dollars 5 cents ; B twice as much ; and C as much as both : what was the whole sum paid ?   Ans. $ 150.30.

QUESTIONS.

1. Why do you set units under units, tens under tens, &c. ?
2. Why do you carry 1 for every ten in the sum of any column ?
3. Why at the last column do you write the whole number ? .
4. Why is Federal Money added and subtracted in the same manner as whole numbers ?
5. What is the use of the decimal point in Federal Money ?
6. Does the operation of *counting* belong to Numeration or Addition ?

SIMPLE SUBTRACTION.

SUBTRACTION is taking a less number from a greater, in order to find the difference or remainder : thus, 4 dollars taken from 6 dollars, the remainder or difference is 2 dollars.

The greater number is called the *minuend*, and the less, the *subtrahend*.

The sign denoting subtraction is a horizontal line at the right of the minuend, as 8—3=5, read, 8 less 3 is equal to 5.

RULE. Set the less number under the greater, units under units, tens under tens, &c.; and beginning at the right, take each figure in the lower number from the one above it, and set down the remainder: but if the lower figure is greater than that above, add ten to the upper figure, then subtract, and carry one to the next lower figure.

PROOF. Add the remainder to the subtrahend, and the sum, if the work is right, will be equal to the minuend.

	(1)	(2)	(3)
Minuend	84736	765249	879647
Subtrahend	52914	439186	64938
Remainder	31822		
Proof	84736		

As we cannot take a larger number from a less, when the lower figure exceeds that above it, it is necessary to increase the upper figure before subtracting. We are therefore directed to *add ten to it*. This, of course, increases the whole minuend, by as much as we add; to balance which, and preserve the relative value of the two numbers, 1 is added to the next lower figure, or (which has the same effect) 1 is taken from the next upper figure.

Adding the remainder to the subtrahend, is merely restoring to the minuend, what was taken from it.

4. Sir Isaac Newton was born in the year 1642; how old was he at his death, which occurred in 1727?
                                        Ans. 85 years.

5. Noah's flood happened in the year of the world 1656; how long from that event to the birth of Christ, in the year 4000?   Ans. 2344 years.

6. Gunpowder was invented in the year 1330 after Christ, and the art of printing in the year 1441: how long from each of those events to the current year 1836?

7. The mariner's compass was invented in Europe in the year 1302; how long since that event to the present year?

8. A man having 586 sheep, wishes to increase his flock to 1250 : how many must he buy ? Ans. 664.

9. What number added to 96+142+312, will make 648 ?
Ans. 98

10. A man having a flock of 576 sheep, sells to three several persons, 50 sheep, 84 sheep, and 96 sheep : how many has he left ? Ans. 346.

SUBTRACTION OF FEDERAL MONEY.

RULE. Place dollars under dollars, cents under cents, &c. ; proceed as in whole numbers, and put the separatrix in the remainder, under the decimal points above.

	(1)	(2)	(3)
	$ c	$ c m	$ c m
From	75. 48	246. 05 6	6327. 86 5
Take	58. 76	218. 60 7	4961.
Ans.	$16. 72		

4. From 36 dollars 5 cents, take 28 dollars 60 cents.
Ans. $ 7.45.

5. From 2 dollars, take 5 dimes and 5 mills.
Ans. $ 1.49½.

6. From 2 dollars, take 7 cents. Ans. $ 1.93.

7. From 4 dollars, take 75 cents. Ans. $ 3.25.

8. Owing my neighbor 60 dollars, I paid him at one time, $9.50 ; at another, $15 ; at another, $30.75, and at another, 62½ cents ; how much was then due him ?
Ans. $ 4.12½.

9. A man paid for two sheep, 7 dollars 50 cents, and for a cow, twice as much lacking 75 cents ; how much did she cost him ? Ans. $ 14.25

QUESTIONS.

What do the terms *Minuend* and *Subtrahend* mean ?
Can a greater number be taken from a less ?
What reason can you assign for your answer ?
What relation has Subtraction to Addition ?
What effect has subtraction on the minuend ?

SIMPLE MULTIPLICATION.

MULTIPLICATION is increasing or repeating the greater of two numbers given, as often as there are units in the less or multiplying number: hence it performs the work of many additions in the most compendious manner.

The signs used to denote multiplication, are ($\times$) or (.); thus $6 \times 2 = 12$, and $(6.2) = 12$ signifies, that the product of 6 by 2 is equal to 12.

MULTIPLICATION TABLE.

1	2	3	4	5	6	7	8	9	10	11	12
2	4	6	8	10	12	14	16	18	20	22	24
3	6	9	12	15	18	21	24	27	30	33	36
4	8	12	16	20	24	28	32	36	40	44	48
5	10	15	20	25	30	35	40	45	50	55	60
6	12	18	24	30	36	42	48	54	60	66	72
7	14	21	28	35	42	49	56	63	70	77	84
8	16	24	32	40	48	56	64	72	80	88	96
9	18	27	36	45	54	63	72	81	90	99	108
10	20	30	40	50	60	70	80	90	100	110	120
11	22	33	44	55	66	77	88	99	110	121	132
12	24	36	48	60	72	84	96	108	120	132	144

I. *If the multiplier does not exceed* 12.

RULE. Multiply each figure in the multiplicand by it, carrying 1 for every 10 as in addition of whole numbers.*

EXAMPLES.

1. What number is equal to 3 times 425 ?

                  425 multiplicand.
                    3 multiplier.
                  ————
                 1275 product.

---

* For the method of proof, see Division.

	(2)	(3)	(4)	(5)
Multiplicand	84635	6432	4345	7085
Multiplier	3	4	5	6
Product				

That Multiplication is a short way of performing several additions, may be shown by setting down the multiplicand as many times as there are units in the multiplier, and adding up the numbers. Thus, in the first example:

$$425$$
$$425$$
$$425$$
_____
$$1275$$

It will be seen that the sum of the addition in this case, is the same as the product of the multiplication.

6. There are 320 rods in a mile; how many rods are there in 7 miles? Ans. 2240.

7. How much will 658 barrels of beef come to, at 8 dollars a barrel? Ans. 4264.

8. There are 9 square feet in a square yard: how many square feet are there in 1297 yards? Ans. 11673.

9. A drover bought 1652 young cattle at an average price of 12 dollars per head: what did they come to? Ans. 19824.

II. *When the multiplier consists of several figures.*

RULE. Set the multiplier under the multiplicand: multiply by each significant figure separately, placing the first figure of each product directly under its own multiplier; the sum of all the products will be the required product.

1. What number is equal to 47 times 465?

Multiplicand	465
Multiplier	47
	3255
	1860
Product	21855

The effect of "placing the first figure of each product exactly under its multiplier," is to increase the product of those figures in the multiplier, which have a local value, according to their distance from the place of units. Thus, in the foregoing example, 4 standing in the place of tens, its product will be ten times as great as if it stood in the place of units ; and by removing each figure one place towards the left, the whole product is increased tenfold. In other words, the 4 is in fact 40, and its product expressed in full, is 18600, which being placed for addition, under that of the 7, will bring the significant figures into the position in which they now stand.

			(3)	(4)
2. Multiplicand	75628		64285	48053
Multiplier	409		64	82
	680652			
	302512			
Product	30931852		4114240	3940346

5. Multiply 96038 by 6007.   Ans. 576900266.
6. Multiply 75649 by 579.   Ans. 43800771.
7. Multiply 29831 by 952,   Ans. 28399112.
8. Multiply 25238 by 1217.   Ans. 30714646.

9. A merchant had 25 bales of cloth, each containing 12 pieces, and each piece consisting of 15 yards : how many yards were there in all ?   Ans. 4500.

10. A clock strikes 78 times in 12 hours, or half a day; how many times does it strike in 365 days ?   Ans. 56940.

III. *When there are ciphers on the right hand of either or both of the factors.*

RULE. Place the significant figures under each other, and multiply them ; and to their product, annex the ciphers standing at the right of the factors.

	(1)
EXAMPLE.	94600
	34000
	3784
	2838
	3216400000

As ciphers have no effect on the result, except, by throwing the significant figures into a higher place, to increase the final product, the reason of the rule is obvious.

2. Multiply 47000 by 120.   .   Ans. **5640000.**

IV. *When the multiplier is equal to the product of two other numbers.*

RULE. Multiply by one of the two numbers, and multiply its product by the other number.

EXAMPLE.

1. Multiply 41364 by 42

42=7×6.

$$
\begin{array}{r}
41364 \\
7 \\
\hline
289548 \text{ product of 7,} \\
6 \\
\hline
1737288 \text{ product of 42.}
\end{array}
$$

It is evident in this example, that the product of 42 is equal to 6 times the product of 7, because 42 is 6 times as many as 7.

2. How many yards of cloth in 29 pieces of 27 yards each?   27=9×3.   Ans. 783.

3. Multiply 649 by 56.

4. Multiply 7028 by 81.

5. Multiply 2586 by 72.

6. Multiply 9407 by 96.

V. *To multiply by* 10, 100, 1000, &c.

RULE. Annex to the multiplicand the ciphers in the multiplier.

1. Multiply 462 by 10.   Ans. 4620.

Annexing a cipher to the right hand of a number, throws every figure one place toward the left, and consequently increases its value ten-fold; that is, *multiplies the whole by* 10. Annexing 2 ciphers removes every figure two places toward the left, and so on.

2. Multiply 568 by 100   Ans. 56800

3. There are 1000 mills in 1 dollar: how many mills in 525 dollars?   Ans. 62500C

4. Multiply one million by one hundred thousand.

VI. *If either of the factors consist of the denominations of Federal Money.*

The separatrix in the product must be as many places from the right, as it is in that factor.

2*

EXAMPLE.

1. What will 15 yards of broadcloth come to, at 5 dollars, 62 cents, and 5 mills, per yard?

$$\begin{array}{r} \text{\$ \ c \ m} \\ 5.625 \\ 15 \\ \hline 28125 \\ 5625 \\ \hline \end{array}$$

Ans. $ 84.375

2. What will 56 lb. of pork come to at 6 cents per lb.?

$$\begin{array}{r} 56 \\ .06 \\ \hline \end{array}$$

Ans. $3.36

3. 56 bushels of potatoes at 33 cents per bu.
Ans. $ 18.48
4. 27 yds. of shirting at 75 cents per yd.   Ans. $ 2 0.25.
5. 48 lb. of cheese at 7 cents per lb.   Ans. $ 3.36.
6. 29 lb. of sugar at 14 cents per lb.   Ans. $ 4.06,
7. 60 lb. of tea at $1.12 cents 5 mills per lb. Ans. $ 67.50.
8. 20 reams of paper at $4.25 per rm.   Ans. $ 85.00.
9. 13 yards of silk at $1.25 per yd.   Ans. $ 16.25.
10. 50 bushels of wheat at 75 cents per bu. Ans. $ 37.50.

QUESTIONS.

Does it make any difference in the product, which of the two given numbers you make the multiplier?

Why is the largest number usually made the multiplicand, and the smallest the multiplier?

Multiplication is but another method of performing addition, is it then indispensable in practice?

Why do we place the product of the second figure in the multiplier, one place to the left of the product of the first figure?

Why in multiplying do we begin with the lowest figure in the multiplier, rather than the highest?

Provided we place the products for addition, in the order of their respective values—does it make any difference in the result, which figure of the multiplier we commence with?

If you annex three ciphers to the right hand of a number, how much is its value increased?

SIMPLE DIVISION.

DIVISION is the operation, by which we find how many times one number is contained in another, and is a concise way of performing several subtractions. Three principal parts are to be noticed :

1. The *Dividend*, or number to be divided.

2. The *Divisor*, or number by which it is divided.

3. The *Quotient*, or result of the division, showing how many times the divisor is contained in the dividend.

When the divisor is not contained in the dividend an *exact* number of times, what is *over* or *remaining* after the division, is thence called the *Remainder;* it must of course, be *less* than the divisor.

The sign commonly used to denote Division, is ÷ thus, 12÷3=4. Division is also indicated by placing the divisor under the dividend with a line between them : thus, $\frac{12}{3}$=4, which is read 12 divided by 3 equals 4.

DIVISION TABLE.

2 in	2	1 time	3 in	3	1 time	4 in	4	1 time
2 in	4	2 times	3 in	6	2 times	4 in	8	2 times
2	6	3	3	9	3	4	12	3
2	8	4	3	12	4	4	16	4
2	10	5	3	15	5	4	20	5
2	12	6	3	18	6	4	24	6
2	14	7	3	21	7	4	28	7
2	16	8	3	24	8	4	32	8
2	18	9	3	27	9	4	36	9
2	20	10	3	30	10	4	40	10
5 in	5	1 time	6 in	6	1 time	7 in	7	1 time
5 in	10	2 times	6 in	12	2 times	7 in	14	2 times
	15	3	6	18	3	7	21	3
	20	4	6	24	4	7	28	4
	25	5	6	30	5	7	35	5
	30	6	6	36	6	7	42	6
	35	7	6	42	7	7	49	7
	40	8	6	48	8	7	56	8
	45	9	6	54	9	7	63	9
5	50	10	6	60	10	7	70	10

8 in	8	1 time	9 in	9	1 time	10 in	10	1 time
8 in	16	2 times	9 in	18	2 times	10 in	20	2 times
8	24	3	9	27	3	10	30	3
8	32	4	9	36	4	10	40	4
8	40	5	9	45	5	10	50	5
8	48	6	9	54	6	10	60	6
8	56	7	9	63	7	10	70	7
8	64	8	9	72	8	10	80	8
8	72	9	9	81	9	10	90	9
8	80	8	9	90	10	10	100	10

11 in	11	1 time	12 in	12	1 time
11 in	22	2 times	12 in	24	2 times
11	33	3	12	36	3
11	44	4	12	48	4
11	55	5	12	60	5
11	66	6	12	72	6
11	77	7	12	84	7
11	88	8	12	96	8
11	99	9	12	108	9
11	110	10	12	120	10

RULE Find how many times the divisor is contained in as many of the left hand figures of the dividend, as are just sufficient to contain it, and set the result on the right of the dividend, for the first quotient figure ; multiply the divisor by it, and place the product under the part of the dividend taken ; subtract it therefrom, and to the right of the remainder, bring down the next figure in the dividend : divide this number as before, and thus proceed until every figure in the dividend is brought down.

PROOF. Multiplication and Division reciprocally prove each other. If the product of two numbers be divided by either factor, the quotient will be the other factor; and if the divisor and quotient of a division be multiplied together, and the remainder—if any—be added, the product will be the dividend. Therefore,

To prove Multiplication.

Divide the product by the multiplier, and the quotient will be the multiplicand.

To prove Division.

Multiply the quotient and divisor together, and add the remainder* to the product, and the result will be the dividend.

1. Divide 9391 by 4.

```
4)9391(2347¾ quotient.
  8
  ——
  13
  12
  ——
  19
  16
  ——
  31
  28
  ——
  3 remainder.
```

2. Divide 1632 by 12.

```
12)1632(136 quotient.
   12                   136
   ——                    12
   43                   ——
   36                  1632 proof.
   ——                   ——
   72
   72
   ——
```

The process of Division may be analyzed as follows:

The first figure in the dividend, (9,) stands in the fourth place, and is therefore thousands, 4 is contained in it 2000 times, with a remainder of 1000.† In order to divide this remainder, we unite to it the next figure,(3,) which, standing in the place of hundreds, make 1300. 4 is contained in 1300, 300 times, with a remainder of 100. The next figure (9) being united to this remainder, makes 190, and 4 is contained in it 40 times with a remainder of 30; and the last figure brought down makes with it 31. 4 is contained in this 7 times, with a remainder of 3 units. The several

```
4)9391(2000 times.
  8000  300  "
  ——     40  "
  1300    7  "
  1200 ——
  ——  2347—3
  190 ——
  160
  ——
   31
   28
   ——
    3
   ——
```

---

* Or subtract it from the dividend, and the remainder will equal the product of the divisor and quotient.

† The quotient figure being thousands, the divisor is not contained but 2 thousand times.

quotients being added up in their proper order, make 2347, the entire quotient.

In the actual division, however, the successive figures of the dividend may all be treated as units, and the ciphers in the several quotients omitted, as their places are to be supplied by the successive quotient figures.

NOTE. The remainder may either be carried out to the right of the quotient, with a line between, or be placed above the line, and the divisor below, in the form of a fraction.*

To know how many times the divisor is contained in the part of the dividend taken at any one division.

If the number of figures in the divisor is *equal* to the number of figures taken in the dividend, divide the first in the dividend by the first in the divisor ; if the number taken in the dividend *exceed* those in the divisor, divide the first two in the dividend by the first in the divisor, making allowance in both cases for what you may have to carry for the multiplication of the left hand figure, but one, of the divisor.

3. Divide 1632 by 136.

```
136)1632(12        845)9461(11
    136       136      845        845
    ———        12      ———         11
    272       ———     1011        ———
    272  Proof 1632    845        9295
    ———       ———     ———         166
                                  ———
              Remainder 166      9461  proof.
    ———       ———     ———
```

4. Divide 74638105 by 37.   Ans. 2017246$\frac{3}{37}$.
5. Divide 4637064283 by 57606.   Ans. 80496$\frac{45787}{57606}$.
6. Divide 72091365 by 5201.   Ans. 13861$\frac{304}{5201}$.
7. If a man's income be 2596 dollars a year ; how much

---

* This fractional expression for a remainder, will be better understood, when the subject of fractions comes to be treated of. When both the terms can be divided, by any number without remainder, it should be reduced in this manner.

is it a week, allowing 52 weeks to the year; and how much a day, there being 7 days in a week.   Ans. $ 48 a week, $ 6$\frac{6}{7}$ a day.

8. If you have a journey to perform of 1260 miles, and travel on an average, 36 miles a day: how many days will be requisite to perform the journey?   Ans. 35 days.

9. In a solid foot there are 1728 cubic inches: how many solid feet in 207360 cubic inches?   Ans. 120.

10. How many hogsheads would be required to contain 2646 gallons of melasses; one hogshead containing 63 gallons.   Ans. 42.

II. *When the divisor does not exceed* 12

The multiplication and subtraction may be made in the mind, without the formality of setting down the whole process. In this case, the quotient is placed under the dividend with a line drawn between.

EXAMPLES:

(1)	(2)	(3)
3)56103961	6)38072940	8)23718620

Quotient 18701320$\frac{1}{3}$

(4)	(5)	(6)
9)43081962	11)57014230	12)27980313

7. A yard is 3 feet: how many yards are there in 5292 feet?   Ans. 1764.

8. At 4 dollars a head, how many sheep can be purchased with 2684 dollars?   Ans. 671.

9. How many yards of cloth, at 7 dollars a yard, can be purchased with 441 dollars?   Ans. 63.

10. If a man's income be 1248 dollars a year; how much is it per calendar month?   Ans. $ 104.

CONTRACTIONS IN DIVISION.

III. *When the divisor is a composite* number, and can be resolved into convenient factors.*

RULE. Divide first by one of those factors, and the quotient thence arising by the other factor : the last quotient will be the answer.

To find the true rem   nder,† multiply the first divisor and last remainder together, and to the product add the first remainder.

1. Divide 182641 by 72.

```
72=9×8      9)182641        First divisor      9
            ────────        Last remainder     5
            8)20293——4                        ──
            ────────                           45
               2536——5      First remainder    4
            ────────                           ──
                            Whole remainder    49
                                               ──
```

Ans. 2536$\frac{49}{72}$

This contraction is the converse of *multiplication by a composite number ;* and is easily understood from the example. The factor 9, will evidently be contained in 182641, eight times as often as 72, the whole divisor. Its quotient, therefore, will be eight times as large as the true quotient. Consequently, dividing by 8 reduces it to an equality with the quotient of 72.

2. Divide 356928 by 32.   Ans. 11154.
3. Divide 934824 by 48.   Ans. 19475$\frac{1}{2}$.
4. Divide 290520 by 108.   Ans. 2690.
5. Divide 2316 by 96.   Ans. 24$\frac{1}{8}$.
6. Divide 2025 by 132.   Ans. 15$\frac{45}{132}$.

---

* A *composite* number, is one which is the exact product of other smaller numbers ; thus, 24=6 × 4.

† True remainder. In the example, the remainder 4, (after the first division,) united to the quotient fractionally, makes the second dividend 20293$\frac{4}{9}$. After the second division, there is, therefore, a remainder of 5$\frac{4}{9}$, which being reduced to an improper fraction, makes $\frac{49}{9}$. This is to be divided by the last divisor (8), and gives us $\frac{49}{72}$.

IV. *To divide by* 10, 100, 1000, &c.

RULE. Cut off as many figures from the right of the dividend, as there are ciphers in the divisor. The figures on the left of the point will be the quotient; and those on the right the remainder.

$$7854 \div \begin{cases} 10 & =785.4 \\ 100 & =78.54 \\ 1000 & =7.854 \end{cases} \qquad \begin{array}{l} \text{Ans. } 785\frac{4}{10}. \\ \text{Ans. } 78\frac{54}{100}. \\ \text{Ans. } 7\frac{854}{1000}. \end{array}$$

Multiplying a number by 10, increases its value tenfold; and dividing it by 10, diminishes it in the same ratio. The former is effected by removing every figure in it, one place *farther* from the unit's place; the latter, by bringing them one place *nearer*.

V. *When there are ciphers on the right hand of the divisor.*

RULE. Cut off the ciphers, and an equal number of places from the right of the dividend: in dividing, omit the figures cut off, and annex those of the dividend to the remainder, if any; otherwise, the figures cut off from the dividend are the remainder.

1. Divide 3704196 by 20.   2. Divide 369183 by 7100.

2|0)370419|6   71|00)3691|83(51$\frac{7083}{7100}$ Ans.
                        355
Ans. 185209$\frac{16}{20}$ ───
                         141
                          71
                        ───
                          70
                        ───

3. Divide 2976435 by 2800.
4. Divide 9400630 by 4700.
5. Divide 6749802 by 9000.
6. Divide 4872036 by 1200.

When the dividend is Federal Money, the operation is the same as in Division of whole numbers, and the only difficulty lies in determining of what denominations the quotient consists.*

When the divisor is a whole number, the quotient may be regarded as consisting wholly of the lowest denomination of the dividend. If that be dollars and cents, the quotient may be considered as so many *cents*.

If there be dollars only, the remainder (if any) may be reduced to cents, by adding 2 ciphers, and continuing the division; or, it may be reduced to mills by adding 3 ciphers.

1. The cost of 61 bushels of barley, being $ 45.75; how much is it a barrel?　　　　　　　　　　　　　Ans. 75 cts.

2. If 15 sheep cost $ 33.75; how much is it a head?

As only *two* places of figures are assigned to *cents*, these are to be separated by a decimal point from those above, which are dollars. If there are *mills* in the dividend, the point falls three places from the right.

```
15)33.75(225 cents, or $ 2.25
   30.
   ____
   37
   30
   ____
   75
   75
   __
```

3. Divide $ 611, equally among 5 men.

To the remainder (1) in this example, two ciphers are annexed for cents.

```
5)611.
  ____
  122.20
  ____
```
　　　　　　　　　　　　　　　　Ans. $ 122.20.

4. Divide $ 171.55 by 47.　　　　　　Ans. $ 3.65.

5. Divide $ 7913.576 by 209.　　　　Ans. $ 37.864.

6. A bunch of quills, consisting of 25, is sold for $ .625; how much is it a piece?　　　　　Ans. 25m. or 2½ cents.

---

* Federal Money will be treated of more at large, under the head of *Decimals*, to which it properly belongs.

QUESTIONS.

What does the quotient in division show ?

Why must the remainder always be less than the divisor ?

Define the three principal parts in division.

How do we know what number of figures in the dividend to take at each step of the operation ?

Division " is a concise way of performing several subtractions :"— work the following question by both methods, and show which is the shortest—

There are in a vessel 17 quarts of water, how many times could a gallon measure be filled full from the vessel ?

Can a less number be divided by a greater ?

If a number be divided by unity, (1) what will be the quotient ?

If a number be divided by less than unity, ($\frac{1}{2}$ for example,) ought the quotient to be more or less than the dividend ?

Of two numbers as divisors, with the *same* dividend, which gives the largest quotient, the greater or the less divisor ?

In proving division, why is the remainder added to the product of the divisor and quotient ?

SUPPLEMENT TO

MULTIPLICATION AND DIVISION.

I. *When the multiplier has a fraction united to it, as* 6$\frac{1}{2}$, 6$\frac{3}{4}$, *&c.*

Take the parts of the multiplicand denoted by the fraction, and add them to the product of the whole number.

Multiplying by 1, gives us the multiplicand for a product ; therefore, multiplying by $\frac{1}{2}$, $\frac{1}{4}$, &c., must give $\frac{1}{2}$, $\frac{1}{4}$, (according as the fraction may be,) of the multiplicand.

1. What will 720 barrels of flour come to, at $ 6$\frac{1}{2}$ per barrel ?

$\frac{1}{2}$)720
    6
———
4320
 360
———
Ans. $ 4680

At $ 6$\frac{3}{4}$

$\frac{1}{4}$)720 = 180
   6   3
———
4320 540
 540 ———
———
Ans. $ 4860

2. 48 men were to receive $ 5⅜ a piece ; how many dollars paid them ?

3. What will 15 tons of hay come to, at $ 7⅔ a ton ?

4. A farm of 156 acres was sold for $ 34⅝ an acre : how much did it come to ?

To divide by a mixed number, as 5½, 5¾, *&c.*

Multiply the whole number by the lower term of the fraction, add the upper term to the product for a divisor ; then multiply the dividend by the lower term of the fraction and divide.

1. There being 5½ yards in a rod, how many rods in 682 yards ?

5×2+1=11   682×2=1364÷11=124 Answer.

2. How many barrels of flour at $ 5¾, can be bought with $ 500 ?   Ans. 86$\frac{22}{23}$.

3. 31½ gallons to the barrel, how many barrels in 485 gallons ?   Ans. 15$\frac{25}{63}$.

II. Operations in arithmetic, may often be considerably shortened, by a little attention to the relations of numbers. A few examples are adduced of contractions in multiplication.

Suppose it were required to multiply a number—say 746382—by 999.

Multiply it by 1000	746382000
Subtract the multiplicand	746382

Ans. 745635618

Here it is evident, that the product of 1000 exceeds the product of 999, by once the multiplicand.

	Multiply 4532	Multiply 4532
	By 639	By 963
63=9×7.	40788	40788
	285516	285516
	Ans. 2895948	Ans. 4364316

In both the foregoing examples, we multiply the product of 9 by 7 : because 63 is equal to 7 times 9. But, because the 9 in the last example, stands in the place of hundreds, the product of the other two figures is set two places toward the right.

(1)

$144 = 12 \times 12$

Multiply 785460
By   14412
———
9425520
113106240
———
Ans. 11320049520

(2)

$49 = 7 \times 7$

Multiply 40788
By   497
———
285516
1998612
———
Ans. 20271636

(3)

$18 = 6 \times 3$

Multiply 576
By 186
———
3456
10368
———
Ans. 107136

(4)

Multiply 576
By 618
———
3456
10368
———
Ans. 355968

It will be perceived, that it is indifferent in what order the figures of the multiplier stand, or which one of them is first multiplied by; the value of the products depending alone on their local position.

Whenever the multiplier has in it a figure, and the multiple* of that figure, this contraction may be used.

The following combinations will exhibit multiples of 6, 126, 186, 246, 306, 366, 426, 486, 546, &c. ; or placing the 6 first : 612, 618, 624, 630, 636, 642, 648, &c.

---

* By *multiple* is meant, the exact product of any number by another.

(3*)

EXAMPLES.

Multiply 8340745 by 64324* and by 64432.
Multiply 61524 by 273 and by 327.
Multiply 342516 by 7209 and by 9072.
Multiply 587632 by 903 and by 309.

III. *When a number is to be multiplied by a second, and that product to be divided by a third, it is often practicable to abridge the operation.*

For example : If it were required to multiply 420 by 18, and divide the product by 6.

$$18 \div 6 = 3 \times 420 = 1260.$$

Here it is quite evident, that multiplying by the sixth part of eighteen, is equivalent to multiplying by the whole number, and taking the sixth part of the product.

Multiply 174 by 48, and divide the product by 12.

$$48 \div 12 = 4 \times 174 = 696.$$

IV. *When the multiplier and the divisor can be reduced proportionally, by dividing both by the same number, the quotients may be substituted for the numbers themselves.*

1. Multiply 81 by 24, and divide the product by 36.

$$24 \div 12 = 2$$
$$36 \div 12 = 3 \qquad 81 \times 2 \div 3 = 54 \text{ Ans.}$$

In the following examples, multiplication and division will be indicated by their appropriate signs.

2. $84 \times 56 \div 14 =$	7. $75 \times 21 \div 7 =$
3. $84 \times 14 \div 56 =$	8. $375 \times 7 \div 21 =$
4. $126 \times 72 \div 48 =$	9. $140 \times 36 \div 84 =$
5. $8094 \times 96 \div 32 =$	10. $4783 \times 39 \div 13 =$
6. $5728 \times 49 \div 56 =$	11. $5204 \times 28 \div 56 =$

---

* Observe that $64 = 32 \times 2$, and $32 = 4 \times 8$.

For convenience in these operations, place the terms of the dividend above a horizontal line, with a sign of multiplication between, and the terms of the divisor beneath; then divide any term above, and any one below the line, by some number which will divide both without remainder; and substitute the quotients for the terms themselves. Thus: ——— ———

12. $132 \times 56 \div 24 \times 7$.

$$\text{Dividend} \quad \frac{132 \times 56}{24 \times 7} = \frac{11 \times 8}{2 \times 1} = \frac{44}{1}, \text{ or } 44 \text{ Ans.}$$
Divisors

Dividing the first two terms by 12, and the other two by 7; and again dividing two of the quotients by 2, we obtain
$\frac{11 \times 4}{1 \times 1}$. If a perpendicular line be used, and the terms set under each other, the sign of multiplication may be omitted.

13. $64 \times 18 \times 48 \div 16 \times 9 \times 12$.

$$
\begin{array}{c|c|c|c}
16 & 64 & 1 & 4 \\
9 & 18 = & 1 & 2 = 32 \text{ Ans.} \\
12 & 48 & 1 & 4 \\
\end{array}
$$

14. $48 \times 25 \div 32 \times 15$.

$$
\begin{array}{c|c|c|c|c}
32 & 48 & 4 & 6 & 2 & 1 \\
15 & 25 = 3 & 5 = 1 & 5 = 2\frac{1}{2} \text{ Ans.} \\
\end{array}
$$

But one line is necessary in the operation, as the quotients may be placed at the right or the left of their respective terms, which should have a pencil drawn across them, to show that they are disposed of. The quotients may again be divided, if any two on opposite sides of the line are divisible by a common number. When the division is finished, the numbers remaining in the dividend are multiplied, and their product divided by the product of those remaining in the divisor.

15. $80 \times 9 \times 21 \div 60 + 6 \times 14$.

$$
\begin{array}{c|c|c}
60 & 80 & 3 & 4 \\
6 & 9 = 2 & 3 = 3 \text{ Ans.} \\
14 & 21 & 2 & 3 \\
\end{array}
$$

Questions to exercise Multiplication and Division.

1. There are 20 pieces of cloth, each containing 27 yards; and 49 other pieces, each containing 75 yards; how many yards are there in both the pieces? Ans. 4215.

2. A person dying, leaves an estate of 4500 dollars to

be divided equally among 5 children : what is the share of
each ?   Ans. $ 900.

3. How many times can 24 be subtracted from 1476 ?
                                        Ans. 61⅓ times.

4. If 2550 lb. of butter are to be packed into firkins,
30 lb. in each ; how many firkins will be required.
                                        Ans. 85.

5. A farmer paid 7950 dollars for a farm ; he sold 50
acres for 40 dollars an acre, and the remainder stood him
in 35 dollars per acre : how much land did he purchase ?
                                        Ans. 220 acres.

6. The product of the divisor and quotient being 800,
the quotient 32, and the remainder 17 : what is the divisor,
and what the dividend ?   Ans. { Divisor 25.
                                       { Dividend 817.

NOTE. See the method of proof for multiplication and
division.

7. $ 17155 prize money were to be divided among 47
seamen : how many dollars had each ? ·   Ans. $ 365.

8. $ 6560 were divided equally among a certain number
of men ; half a share (amounting to $ 41) was paid down,
and the residue in 3 months : how many men were there ?
                                        Ans. 80 men.

9. A field is to be fenced, two sides of which are each
63 rods in length, and the other two are each 42 rods ;—
allowing 9 rails to a rod, how many will be required
to fence the field ? and if there are 7 rails to a length, how
many lengths will there be ?   Ans. { 1890 rails.
                                       { 270 lengths

10. The product of two numbers is 9594, and the multiplicand 246 ; what is the multiplier ?   Ans. 39.

11. The number of sheep in two separate flocks is 340,
the larger flock has 40 more sheep in it than the other :
how many sheep are there in each flock ?   Ans. { 190.
                                                { 150.

12. If a quantity of provision will last 324 men 7 days ;
how many men will it last one day ?   Ans. 2268.

13. A garrison of 527 men have provision sufficient to
last 47 days, if each man is allowed 15 oz. per day ; how
many men would it serve the same time, if each man were
allowed 5 oz. a day ?   Ans. 131.

14. If a man performs a journey in 58 days, by travelling 9 hours a day : how many hours is he performing it ?

Ans. 522

15. If by working 12 hours in a day, a man can perform a piece of work in 217 days ; how long would it take him to do it if he worked only 4 hours in a day ? Ans. 651.

16. At 5 dollars a ream, how many reams of paper may be bought for 253 dollars ? Ans. $50\frac{3}{5}$.

17. At 7 dollars a barrel, how many barrels of flour may be bought for 2434 dollars ? Ans. $347\frac{5}{7}$.

18. At 9 dollars a barrel, how many barrels of beef may be bought for 3827 dollars ? Ans. $425\frac{2}{9}$.

19. At the rate of 150 miles per day, how many days will it take a ship to make a voyage of 3000 miles ?

Ans. 20.

20. If 64 gallons of water in one hour, run into a cistern, containing 768 gallons, in how many hours will it be filled ?

Ans. 12.

21. At 8 dollars a cord, how many cords of wood may be bought for 853 dollars ? Ans. $106\frac{5}{8}$.

22. How much indigo at 2 dollars per pound, must be given for 19 yards of broadcloth, at 7 dollars per yard.

Ans. $66\frac{1}{2}$.

23. How many bushels of corn, at 5 shillings per bushel, must be given for 23 bushels of wheat, at 7 shillings per bushel ? Ans. $32\frac{1}{5}$.

COMPOUND NUMBERS.

COMPOUND NUMBERS are such as express quantities consisting of different denominations, but of the same generic kind ; as Tons, Hundreds, Quarters, &c. ; Bushels, Pecks, Quarts, Pints.

The following Tables of the denominations of Compound Numbers, are to be thoroughly committed to memory before entering upon reduction.

STERLING MONEY.

The denominations of English Money are, the pound, £. ; the shilling, s. ; the penny, d. ; and the farthing, qr.

4 farthings	= 1 d.	1 qr.	= $\frac{1}{4}$ of 1 d.
12 pence	= 1 s.	1 d.	= $\frac{1}{12}$ of 1 s.
20 shillings	= 1 £.	1 s.	= $\frac{1}{20}$ of 1 £.

In the rest of the tables, let the pupil form for himself the fractional part, which each inferior denomination is of the next superior.

TROY WEIGHT.*

The denominations of Troy Weight are, the pound, lb.; the ounce, oz. ; the pennyweight, dwt. ; and the grain, gr.

24 grains	- - -	= 1 dwt.
20 pennyweights	-	= 1 oz.
12 ounces	- - -	= 1 lb.

By this weight are weighed gold, silver, and jewels.

APOTHECARIES' WEIGHT.

The denominations of Apothecaries' Weight are, the pound, ℔ ; the ounce, ℥ ; the dram, ℨ ; the scruple, ℈ ; and the grain, gr.

20 grains	- - -	= 1 ℈,
3 scruples	- - -	= 1 ℨ,
8 drams	- - -	= 1 ℥,
12 ounces	. - -	= 1 ℔.

---

* " The original of all weights used in England, was a *grain*, or corn of wheat, gathered out of the middle of the ear and being well dried, 32 of these were to make one *penny-weight*, 20 penny-weights one ounce, and 12 ounces one pound. But in later times, it was thought sufficient ⁻∩ divide the same penny-weight into 24 equal parts, still called **grains**, ɴg the least weight now in common use."—*Dr. Hutton.*

This, like Troy Weight, has for its basis, *grains,* with some different divisions in the higher denominations. It is used by apothecaries in compounding medicines.

AVOIRDUPOIS WEIGHT.

The denominations of Avoirdupois Weight are, the ton T.; the hundred-weight, cwt.; the quarter, qr.; the pound, lb.; the ounce, oz.; and the dram, dr.

16 drams - - - =	1 oz.
16 ounces - - - =	1 lb.
28 pounds - - - =	1 qr.
4 quarters - - - =	1 cwt.
20 hundred-weight =	1 T.

By this are weighed all metals, except gold and silver, and heavy and drossy articles in general.

The whole weight of commodities weighed by Avoirdupois, including the box, cask, &c., containing them, is called *gross weight;* and what remains, after allowance for the weight of the box, cask, &c., and for probable waste in some kinds of articles, is called *neat weight.*

In avoirdupois, a hundred weight is equal to 112 lb. In the state of New York, however, it is fixed by statute at 100 lb., and the quarter at 25 lb. The fitness and convenience of such a division, commends it to general adoption.

CLOTH MEASURE.

The denominations of Cloth Measure are, the French ell, Fr. e.; the English ell, E. e.; the Flemish ell, Fl. e.; the yard, yd.; the quarter, qr.; and the nail, na.

4 nails - - - - =	1 qr.
4 quarters - - - =	1 yd.
3 quarters - - - =	1 Fl. e.
5 quarters - - - =	1 E. e.
6 quarters - - - =	1 Fr. e

DRY MEASURE.

The denominations of Dry Measure are, the bushel, bu. ;
the peck, pk. ; the gallon, gal. ; the quart, qt. ; the pint, pt.

2 pints - - - - - - -	= 1 qt.
4 quarts - - - - - -	= 1 gal.
8 quarts - - - - - -	= 1 pk.
4 pecks - - - - - -	= 1 bu.*

By this are measured grains, seeds, fruits, salt, &c.

WINE MEASURE.

The denominations of Wine Measure, are, the tun, T. ;
the pipe, p. ; the puncheon, pun. ; the hogshead, hhd. ; the
tierce, tier. ; the barrel, bl. ; the gallon, gal. ; the quart, qt. ;
the pint, pt. ; the gill, gi.

4 gills - - -	= 1 pt.
2 pints - - -	= 1 qt.
4 quarts - - -	= 1 gal.
31½ gallons - - -	= 1 bl.
42 gallons - - -	= 1 tier.
63 gallons - - -	= 1 hhd.
84 gallons - - -	= 1 pun.
126 gallons - - -	= 1 p.
2 pipes - - -	= 1 T.

By this are measured wines, spirits, cider, oil, &c.

BEER MEASURE.

The denominations of Beer Measure are, the butt, bt :
the hogshead, hhd. ; the barrel, bl. ; the kilderkin, kil. ;
the firkin, fir. ; the gallon, gal. ; the quart, qt. ; and the
pint, pt.

---

* 8 bushels - - - -	= 1 quarter English,
36 bushels - - - -	= 1 chaldron of coal.

2 pints - - - -	=	1 qt.
4 quarts - - - -	=	1 gal.
9 gallons - - -	=	1 fir.
2 firkins - - - -	=	1 kil.
2 kilderkins - - -	=	1 bl.
3 kilderkins - - -	=	1 hhd.
2 hogsheads - -	=	1 bt.

Note. The dry gallon contains $268\frac{4}{5}$ cubic inches; the wine gallon, 231 cubic inches, and the beer gallon 282 cubic inches.

LONG MEASURE.

The denominations of Long Measure are, the mile, m.; the furlong, fur.; the rod, or pole, r.; the yard, yd.; the foot, ft., and the inch, in.

12 inches - - -	=	1 ft.
3 feet - - - -	=	1 yd.
$5\frac{1}{2}$ yards - - -	=	1 r.
40 rods - - - -	=	1 fur.
8 furlongs - - -	=	1 m.
3 miles - - - -	=	1 league.

SQUARE MEASURE.

The denominations of Square Measure are, the mile, m.; the acre, A.; the rood, R.; the rod. r.; the yard, yd.; the foot, ft., and the inch, in.

144 inches - - -	=	1 ft.
9 feet - - - -	=	1 yd.
$30\frac{1}{4}$ yards - - -	=	1 r.
40 rods - - - -	=	1 R.
4 roods - - -	=	1 A.
640 acres - - -	=	1 m.

This is applied to the measurement of surfaces, such as land, boards, flooring, &c., where the dimensions of length and breadth only are concerned.

(4)

CUBIC MEASURE.

The denominations of Cubic Measure are, the yard, yd.; the foot, ft., and the inch, in.

1728 inches - - -	=	1 ft.
27 feet - - - -	=	1 yd
40 feet of round timber, or 50 feet of hewn timber, make 1 Ton.		
128 feet, i.e. 8 ft. long, 4 ft. high, and 4 ft. wide.	=	1 cord.

Cubic Measure is applied to solids, which have the three dimensions of length, breadth, and thickness; likewise, to the measurement of capacities, as of cisterns and containers generally.

TIME.

The denominations of Time are, the year, Y.; the week, wk.; the day, d.; the hour, h.; the minute, m., and the second, s.

60 seconds - - -	=	1 m.
60 minutes - - -	=	1 h.
24 hours - - -	=	1 d.
7 days - - -	=	1 wk.

The exact solar year is 365 days, 5 hours, 48 minutes, and 48 seconds. To avoid these fractional parts of a day, the year is made to consist of 365 days; except every fourth year, which has 366.* When the number of the year is divisible by 4 without a remainder, it is Leap-Year— the centurial year excepted.

DIVISION OF THE CIRCLE.

The Divisions of the Circle, C., are, the sign, S.; the degree, (°); the minute, ('); and the second, (").

60 seconds - - -	=	1 '
60 minutes - - -	=	1 °
30 degrees - - -	=	1 S.
12 signs - - -	=	1 C.

---

* Called Leap-Year.

REDUCTION.

REDUCTION is bringing or changing numbers from one denomination to another, without altering their value. Thus, 1 yard, reduced to the next inferior denomination, is 3 feet; and these reduced again, are 36 inches.

Again, lower denominations may be brought to higher; as pounds to quarters, hundred weights, or tons. In all these cases, the quantity remains unchanged, although it is expressed in a different denomination.

When higher denominations are to be reduced to lower.

RULE. Multiply the successive denominations, commencing with the highest given, by as many of the next lower, as make one of that higher, adding in their proper places, the several inferior denominations expressed in the given number.

When lower denominations are to be reduced to higher.

RULE. Divide the given denomination by as many as make 1 of the next higher; and the quotient thence arising, by as many as make 1 of the denomination next above that, and so on to the required one.

The several remainders will be of the denomination of the respective dividends.

1. In 1234*l.* 15*s.* 7*d.*, how many farthings ?

$$\begin{array}{rl}
\pounds \quad s. \quad d. & \\
1234 \;\; 15 \;\; 7 & \\
20 & \\
\hline
24695 & \text{shillings,} \\
12 & \\
\hline
296347 & \text{pence,} \\
4 & \\
\hline
\end{array}$$

Ans. 1185388 farthings.

The reason of the rule hardly requires explanation. To apply it, for instance, to the first example; £1 is equal to 20s. In £1234, then, there will be 20 times as many shillings, as there are pounds; to which, of course, must be added the 15s. in the question.

2. Reduce 1185388 farthings to pence, shillings, and pounds.

$$4)1185388$$

$$12)296347d.$$

$$20)24695s.—7d.$$

Ans. 1234l.—15s. 7d.

3. Reduce 23l. to farthings.   Ans. 22080.
4. Reduce 337587 farthings to pounds.
                     Ans. 351l. 13s. 0¾d.
5. How many farthings in 35 guineas ?*
                          Ans. 35280.
6. In 19l. 16s. 3d. how many shillings, threepences, and farthings ?   Ans. 1585 threepences.
7. In 15l. 2s. how many dollars, at 8 shillings to the dollar ?
8. In $ 46, at 6s. each, how many pounds, and halfpence ?
            Ans. $\begin{cases} £13\frac{16}{20}. \\ 6624 \text{ halfpence.} \end{cases}$

9. In 75 English crowns,† how many pounds, and how many 4½d. ?
            Ans. $\begin{cases} £18\frac{3}{4}. \\ 1000 \text{ } 4½. \end{cases}$

10. In 59 lb. 13 dwt. 5 gr. how many grains ?
                          Ans. 340157.
11. In 8012131 grains, how many pounds ?
            Ans. 1390 lb. 11 oz. 18 dwt. 19 gr.
12. In 34 E. e. 4 qr., how many Ells Flemish?

---

* A guinea is 21 shillings.   † An English crown is 5 shillings.

13 In 2046 ft. how many yards and rods?

Ans. $\begin{cases} 682 \text{ yd.} \\ 124 \text{ r.} \end{cases}$

14. How many seconds in a lunar month; or, 29 days, 12 hours, 44 minutes, 3 seconds?

Ans. 2551443.

15. What will a bushel of clover seed come to, at the rate of 12½ cents a pint?   •   Ans. $ 8.

. 16. Suppose a hogshead of molasses, which cost $ 23, to be retailed at 12½ cents a quart, what is the profit on it?

Ans. $ 8.50.

17. How many times will a wheel, 16 feet 6 inches in circumference, turn round in going 100 miles?

Ans. 32000.

18. How many bottles, containing a pint, a quart, and 2 quarts, an equal number of each, can be filled from half a pipe of wine?   '   Ans. 72 of each.

19. If a barrel of beer be retailed at 3 cents the half pint, how much will it amount to?   Ans. $ 17.28.

20. In 20571005 drams, how many tons?

Ans. 35 T., 17 cwt., 1 qr., 23 lb., 7 oz., 13 dr.

21. In 469 dollars, how many cents and mills?

Ans. 46900 cts., 469000 m.

NOTE. As 100 cents and 1000 mills, each make 1 dollar, the reduction of dollars to cents, is performed by simply annexing 2 ciphers, and to mills by annexing 3 ciphers.

If the number be dollars and cents, it is reduced to cents by erasing or disregarding the decimal point. So, likewise, if it contain mills, the whole may be considered as mills.

Mills are reduced to dollars by cutting off three, and cents by cutting off two, of the right hand figures by a separatrix; in other words, *dividing by* 1000, *and by* 100.

22. How many cents in $ 65?   Ans. 6500.

23. How many mills in $ 17?   Ans. 17000.

24. Reduce 12 dollars 5 cents to cents.   Ans. 1205.

25. Reduce 9 dollars, 6 cents, 5 mills, to mills.

Ans. 9065.

26. Reduce 15 cents, to mills.   Ans. 150.

27. How many dollars in 1246 cents?   Ans. $ 12.46.

28. How many dollars in 1246 mills?   Ans. $ 1 24 6

(4*)

29. How many mills in 5 pieces of the gold coin denominated eagles?   Ans. 50000.

30. If you purchase 75 bunches of quills, each containing 25 quills, at the rate of 6 mills a piece, how many dolars will they cost?   Ans. $ 11.25.

QUESTIONS.

Is the value of any given number changed by reduction to another denomination?

How do you know when to employ multiplication, and when division, in performing numbers?

One of these methods has sometimes been denominated " *Reduction Ascending*," and the other, " *Reduction Descending*;" you can probably designate both of them.

If you wished to *prove* the accuracy of your work, what method would you take?

When in Reduction Ascending you have a *remainder*, after any division, of what denomination is it?

Of what denomination is the *quotient* in relation to the *dividend?*

ADDITION OF COMPOUND NUMBERS.

RULE. Place the numbers so that the same denominations may stand under each other; add the several columns containing the different denominations, separately, beginning with the lowest; and divide the sum of each by the number of that denomination, which makes one of the next higher; set down the remainder, and carry the quotient to the next column.

	(1)					(2)		
£	s.	d.	qr.		£	s.	d.	qr.
.104	12	6	2		176	6	9	1
50	9	1	0		204	11	0	3
3	18	0	3.		183	19	11	2
Sum 158	19	8	1					

· The propriety of a distinct rule for the addition of compound numbers will be evident, if we reflect that pounds, shillings, pence, &c., cannot be put into one common aggregate, which would express the amount in any one of the denominations.

3. What is the sum of £16 8s. 4d., £19 12s. 7d. 3qr. £104 15s. 11d. 1qr., £20. 11d. 1qr.?

4. Add together, 4 lb. 10 oz. 17 dwt. 13 gr., 1 lb. 9 oz. 8 dwt. 7 gr., 11 oz. 15 dwt. 10 gr., 1 lb. 6 dwt., Troy.

5. What is the sum of 2 oz. 15 dwt. 11 gr., 1 lb. 9 oz. 15 dwt., and 19 dwt. 14 gr. of silver?

6. Add together 16 cwt. 2 qr. 11 lb. 14 oz., 18 cwt. 3 qr. 22 lb. 8 oz., 3 qr. 9 lb. 11 oz., and 17 cwt. 1 qr. 13 oz.

7. What is the weight of 4 loads of hay, weighing severally, 1 T. 3 cwt. 20 lb., 19 cwt. 3 qr. 16 lb., 1 T. 5 cwt., and 18 cwt. 2 qr. 14 lb.?

8. How much is 16 yd. 3 qr. 3 n., 12 yd. 1 qr. 2 n., 9 yd. 3 n., and 6 yd. 2 qr. 1 n. of cloth?

9. What is the sum of 4 E. e. 3 qr. 2 n., 7 E. e. 4 qr. 3 n., 5 E. e. 3 qr., and 4 qr. 3 n. of cloth?

10. How much corn in 3 bags, containing severally 3 bu. 2 pk. 7 qt., 2 bu. 1 pk. 5 qt., and 2 bu. 6 qt.

11. Add together 2 hhd. 51 gal. 2 qt. 1 pt., 3 hhd. 10 gal 1 pt., 60 gal. 3 qt., and 1 hhd. 3 qt. 1 pt.

12. How much wine in 4 p. 125 gal. 3 qt., 68 gal. 2 qt. 1 pt., 34 p. 59 gal., 43 gal. 3 qt. 1 pt.?

13. Add 58 m. 5 fur. 23 r., 17 m. 4 fur. 18 r., 23 m. 39 r., 16 m. 7 fur. 1 r., 6 fur. 30 r., and 20 m. 1 fur. 12 r.

14. Add 46 m. 6 fur. 29 r. 15 ft. 10 in., 39 m. 1 fur. 36 r. 14 ft. 6 in., 53 m. 7 fur. 24 r. 9 ft. 8 in., 7 m. 39 r. 11 ft. 9 in.

15. How much land in 4 lots, containing severally 36 A. 2 R. 18 r., 59 A. 3 R. 12 r., 75 A. 39 r., and 19 A. 11 R. 39 r.?

16. Add 4 yd. 5 ft. 68 in., 3 yd. 8 ft. 114 in., 11 yd. 2 ft. 58 in., and 6 yd. 7 ft. 108 in. square measure.

SUBTRACTION OF COMPOUND NUMBERS.

RULE. Place the numbers as in Compound Addition. Begin with the lowest denomination in the subtrahend, and take it from the number above it ; but if that above be the least, add as many to it as make one of the next higher denomination ; then subtract, and carry 1 to the next denomination in the subtrahend.

The method of proof is the same as in simple subtraction

	(1)				(2)				(3)		
	£	s.	d.	qr.	£	s.	d.	qr.	£	s.	d.
From	24	11	8	3	85	15	10	1	36	9	7
Take	19	15	7	1	63	18	9	3	14	6	9
Ans.	4	16	1	2							
Proof	24	11	8	3							

4. Subtract 2 oz. 14 dwt. 23 gr., from 4 oz. 19 dwt. 21 gr.

5. From 4 lb. 9 oz. 16 gr., take 3 lb. 10 oz. 16 dwt. 15 gr.

6. A merchant having received 7 cwt. 2 qr. 14 lb. of sugar, sells 5 cwt. 3 qr. 25 lb. ; how much remains on hand ?

7. Subtract 10 T. 11 cwt. 20 lb. 10 oz. 11 dr., from 13 T. 9 cwt. 1 qr. 12 oz.

8. From 35 ℔. 7 ℥. 3 ʒ. 1 Ɔ. 14 gr., take 17 ℔. 10 ℥. 6 ʒ. 1 Ɔ. 18 gr.

9. If from a piece of cloth containing 20 yd., two garments be taken, each 3 yd. 3 qr. ; how much remains of the piece ?

10. Subtract 15 m. 6 fur. 26 r. 12 ft., from 28 m. 5 fur. 16 r.

11. From a pipe of wine, a merchant sold to one man, 31 gal. 2 qt., to another, 5 gal., and to a third, 3 gal. 1 qt. 2 pt. ; how much remains of the pipe ?

12. From 1 bl. take 16 gal. 3 qt.

13. From 13 bu. 7 qt. 1 pt., take 7 bu. 3 qt. 1 pt.

14. Subtract 75 d. 12 h. 35 m., from 114 d. 9 h. 26 m.

15. From 7 d. 21 h. 59 m. 16 s., take 4 d. 23 h. 41 m. 50 s.

16. From a pile of wood, containing 19 cords 96 ft. were sold at different times, 3 cords 26 ft., 4 cords 19 ft., and 7 cords 28 ft.; how much remains of the pile?

17. What is the difference in longitude between two places, one of which is 75 ° 15 ' 36 " west from an assumed meridian; and the other 46 ° 20 ' 32 " west?

18. From 112 A. 3 R. 25 r. of land, take 19 A. 2 R. 37 r.

MULTIPLICATION OF COMPOUND NUMBERS.

I. *When the multiplicand consists of different denominations.*

RULE. Place the multiplier under the least denomination of the multiplicand; multiply each denomination separately, beginning with the lowest, and observe the same rules for carrying, &c., as in addition of compound numbers.

1. What is the cost of 8 lb. of tea at 5s. 8½d. per lb.?

(1)				(2)			
s	d.	qr.		£	s.	d.	qr.
5	8	2		1	0	9	2
		8					5
Ans. £2	5	8	0	£5	3	11	2

3. 9 cwt. of cheese, at £1 11s. 5d. per cwt.
Ans. £14 2s. 9d.

4. 12 cwt. of sugar, at £3 7s. 4d. per cwt.
Ans. £40 8s.

5. The weight of 3 chests of tea, each weighing 3 cwt. 2 qr. 9 lb.
Ans. 10 cwt. 2 qr. 27 lb.

6. What quantity of wood in 6 parcels, each containing 3 cords and 96 feet?   Ans. 22½ cords

	(7)				(8)				(9)		
	lb.	oz.	dwt.	gr.	lb.	oz.	dr.	sc	yd.	qr.	n.
Multiply	21	1	7	13	2	4	2	1	8	3	2
				4				7			12

	(10)				(11)			(12)			
	gal.	qt.	pt.	gi.	yd.	ft.	in.	bu.	pk.	qt.	pt.
	3	2	1	3	21	2	9	36	3	6	1
				9			11				8

II. *When the multiplier can be resolved into convenient factors.*

Multiply first by one of them, and its product by the other.

1. 15 cwt. of cheese, at 17s. 6d. per cwt.

$15 = 3 \times 5$

```
        17   6
             3
      ----------
       2  12   6
             5
      ----------
Ans. £13   2   6
      ----------
```

2. 63 bushels of oats, at 2s. 3d. per bushel.
        Ans. £7 1s. 9d.

3. How many bushels of oats will feed 21 horses for a week, allowing each horse 4 bu. 3 pk. ?

4. How many yards of cloth in 35 pieces, each containing 27 yd. 3 qr. 2 n. ?   Ans. 975 yd. 2 qr. 2 n.

III. *When no two numbers multiplied together, will exactly make the multiplier.*

Multiply by any two whose product will come the nearest; then multiply the multiplicand by the difference of the product of the two numbers, and the whole multiplier, and either add or subtract the result, as the case may require.

1. 26 yards of cloth, at 3s. 5d. per yard.

$$26 = 5 \times 5 + 1$$

	s.	d.
	3	5
		5
	17	1
		5
4	5	5
3	5	

Ans. £4 8 10

2. 39 cwt. of sugar at £3 10s. 6d. per cwt.

$$39 = 10 \times 4 - 1$$

Ans. £137 9s. 6d.

3. If a man drink 1 pint 2 gills of ale, for 29 successive days; what quantity will he have drank in all?

Ans. 5 gal. 1 qt. 1½ pt.

4. If a soldier's ration of bread be 5 lb. 6 oz. 8 dr. per week; what will it amount to in 52 weeks?

Ans. 2 cwt. 2 qr. 1 lb. 2 oz.

5. 94 casks of cider, at 12s. 2d. per cask.

Ans. £57 3s. 8d.

6. 34 pieces of cloth, each containing 27 yd. 3 qr.

Ans. 943 yd. 2 qr.

7. A merchant bought 95 pairs of shoes at 4s. 6d. 1 qr. a pair; how much did he pay for the whole?

8. A gentleman bought 43 silver spoons, each weighing 2 oz. 14 dwt. 6 gr.; what was the weight of the whole?

DIVISION OF COMPOUND NUMBERS.

RULE. Divide each denomination separately, beginning with the highest; and if there be a remainder, reduce 1 to the next lower denomination, uniting to it the corresponding denomination in the dividend: divide again, and proceed in the same manner to the end.

1. Divide £60 14s. 8d. among 8 men.

£	s.	d.		£	s.	d.
8)60	14	8		7	11	10
						8
Ans. £7	11	10		Proof £60	14	8

NOTE. Let the pupil prove all the examples by multiplication.

2. Divide £28 2s. 1¼d. by 6.   Ans. £4 13s. 8¼d.
3. Divide £135 10s. 6d. by 9.   Ans. £15 1s. 2d.
4. Divide £1332 11s. 8½, by 12.   Ans. £111 0s. 11¼d.
5. If 23 lb. 4 oz. 6 dwt. 10 gr. of silver, be made into 7 tankards of equal weight, what will be the weight of each?
              Ans. 3 lb. 4 oz. 0 dwt. 22 gr.
6. If 5 horses consume in a week, 14 bu. 2 pk. 6 qt. of oats, how much is it to a horse?   Ans. 2 bu. 3 pk. 6 qt.
7. If 59 cwt. 3 qr. of sugar be put into 6 hogsheads, how much will each hogshead contain? Ans. 9 cwt. 3 qr. 23½ lb.
8. If £2 12s. 6d. be paid for 5 yards of cloth, how much is it per yard?   Ans. 10s. 6d.

When the divisor is a composite number it may be resolved into factors, and the quotient of one be divided by the other.

1. If 16 cwt. of cheese, cost £30 18s. 8d., what is that per cwt.?

	£	s.	d.
	4)30	18	8
16=4×4	4)7	14	8
	Ans. £1	18	8

2. Divide 1061 cwt. 2 qr., by 28.

Ans. 37 cwt. 3 qr. 18 lb.

3. If 35 pieces of cloth, of equal quantity, contain 971 yd. 1 qr., how many yards in a piece ?   Ans. 27 yd. 3 qr.

4. If 259 A. 1 R. 10 r. of land, be divided into 18 equal lots, how much land will be contained in a lot ?

Ans. 14 A. 1 R. 25 r.

5. If 56 lb. of butter, cost £2 9s., what is it per lb. ?

Ans. 10¼d.

6. Divide £124 5s. 4d., into 64 equal parts.

Ans. £1 18s. 10d.

7. Divide 336 bu. 3 pk. 4 qt., by 70.

Ans. 4 bu. 3 pk. 2 qt.

When the divisor is not a composite number, divide by the whole divisor at once, after the manner of long division.

1. Divide 571 yd. 2 qr. 1 n., by 47.

47)571 2 1(12 0 2+* Ans.
47

———
101
94

Here are 7 yards remain-   7
ing, which being reduced to   4
the next denomination, and ———
the 2 added, make 30 quar- 30
ters. This being less than   4
the divisor, has to be reduced ———
to nails ; but a cipher is placed 121
in the quotient, to denote the 94
absence of quarters. ———
    27

2. A merchant bought 109 hogsheads of sugar, weighing 60 T. 17 cwt. 3 qr. 18 lb.; what was the weight of each hogshead ?   Ans, 11 cwt. 0 qr. 19 lb. 6 oz.+

———

* The sign + at the end of the quotient, denotes that the exact answer is somewhat more, which cannot be conveniently expressed, or is too inconsiderable to be regarded. The remainder here, might have been set above the divisor in the usual way, making $\frac{27}{47}$ of a nail.

VULGAR FRACTIONS.

A FRACTION, or broken number, is an expression of some part or parts, of any thing or number considered as a whole.

Fractions are either *Vulgar* or *Decimal.*

A vulgar fraction is denoted by two numbers placed one below the other, with a line between them.

The number below the line is called the *denominator;* and that above the *numerator.*

$$\text{Thus} \quad \frac{5}{8} \quad \begin{array}{l} \text{numerator,} \\ \text{denominator.} \end{array}$$

Both these numbers are also, in general, named the *terms* of the fraction.

Any thing or any number may be conceived to be divided into equal parts. An apple, for example, may be divided into 2, 3, 5, or any other number of equal parts.

If the thing or number be divided into 2 equal parts, each of those parts is called a half; if into 3 equal parts, each of the parts is called a third; if into 4 equal parts, each of the parts is called a fourth; if into 10 equal parts, each of the parts is called a tenth; and so on, into whatever number of parts a thing or number may be conceived to be divided.

It is an axiom, that " the whole is equal to the sum of all its parts ;" and every one will perceive, that 2 halves, 3 thirds, 4 fourths, 5 fifths, 10 tenths, make up the whole of any thing or any number. Thus, there are 2 halves to an apple, 6 sixths, 8 eights, and so on.

Now, fractions are " expressions" for all such parts as any thing or any number can be divided into. If, for example, the division be into halves, it is written, or " expressed" thus, $\frac{1}{2}$; if into thirds, thus, $\frac{1}{3}$, $\frac{2}{3}$; if into sixths, $\frac{1}{6}$, $\frac{2}{6}$, $\frac{5}{6}$, and so on, for any other division into equal parts.

The number below the line, is called denominator because it *gives name* to the fraction, or designates the number of parts, into which the thing or number (otherwise called the integer) is divided.

The number above the line, is called the numerator, because it expresses or shows how many of the equal parts are taken, or used in the fraction. For example, $\frac{5}{8}$ of an apple ; the denominator (8) shows that the apple, (or integer) is divided into 8 equal parts ; and the numerator (5), that 5 of those parts are taken.

A *proper fraction*, is one whose numerator is not greater than its denominator ; as $\frac{2}{13}$, $\frac{4}{5}$.

An *improper fraction*, is one whose numerator exceeds its denominator ; as $\frac{5}{4}$.*

A number consisting of an integer with a fraction annexed, as $14\frac{7}{8}$, is called a *mixed number*.

A *simple fraction*, is a single expression, having but one numerator and one denominator.

A *compound fraction*, is a fraction of a fraction ; as $\frac{3}{4}$ of $\frac{1}{2}$, $\frac{4}{5}$ of $\frac{7}{13}$ of $\frac{3}{2}$.

A whole number may be expressed fractionally, by putting 1 for its denominator ; thus, 4 is $\frac{4}{1}$.

The value of a fraction is equal to the quotient obtained by dividing the numerator by the denominator ; thus, $\frac{9}{3}=3$, and $\frac{20}{5}=4$. Hence, if the numerator be less than the denominator, the value of the fraction is less than 1. If the numerator be greater than the denominator, the value of the fraction is greater than 1.

REDUCTION OF VULGAR FRACTIONS.

Reduction of Vulgar Fractions, is bringing them out of one form or denomination to another, without changing their value.

Whenever the number of parts expressed in a fraction is such, that both the numerator and denominator may be di-

---

* An improper fraction may be regarded as denoting either an *unexecuted* division ; or the division of *more than one* integer into equal parts. Thus to apply the expression $\frac{5}{4}$ to money ;—it may mean that 5 dollars are to be divided into 4 equal parts, or that some number of dollars— more than 1— *being* divided into fourths, 5 of those parts are intended Considered in either way, the expression is equivalent : for $\frac{1}{4}$ of 5 dol·lars, is exactly equal to $\frac{5}{4}$ of 1 dollar. As, however, there can be but $\frac{4}{4}$, $\frac{5}{5}$, &c., to any one integer, such fractions as that above are for distinction denominated *improper* fractions.

vided by 2, 3, 4, or any other number, without a remainder, the fraction may be reduced to fewer, that is larger parts.

I. *To reduce fractions to their lowest terms.*

RULE. Divide the terms of the given fraction by any number which will divide them without a remainder, and the quotients again in the same manner; and so on, till it appears that there is no number greater than 1, which will divide them, and the fraction will be in its lowest terms.*

Or, divide the greater term by the less, and that divisor by the remainder, and so on until nothing remains, then the terms being both divided by the last divisor, will be reduced to the lowest expression.

1. Reduce $\frac{144}{240}$ to its lowest terms.

$$12)\frac{144}{240}=\frac{12}{20}\div 4=\frac{3}{5} \text{ Ans.}$$

---

* As there will be frequent occasion for reducing numbers, other than fractions, by this method, it is important to be able to know, *without trial*, what numbers to adopt as divisors.

1. Any number ending with a cipher, or an even number is divisible by 2.

2. Any one ending with 5 or 0 is divisible by 5.

3. If a number end with 0, it is divisible by 10 ; if with 00, it is divisible by 100, and so on.

4. If the two right hand figures be divisible by 4, the whole is divisible by 4. And if the three right hand figures are divisible by 8, the whole are so.

5. If the sum of the digits in any number be divisible by 3 or 9, the number is divisible by 3 or 9.

6. If the right hand digit be even, and the sum of all the digits be divisible by 6 ; the whole is divisible by 6.

7. Numbers consisting of two and three places, are divisible by 7, when the left hand figure or figures, are double the right hand figure ; thus, 63, 84, 126, 159, 168 ; and when the two right hand figures are five times the left hand figures ; as 525, 630, 735, 840, 945.

8. When the two right hand figures of a number consisting of four places, are five times the two left hand figures, or the two left hand figures *three* times the two right hand, the number is divisible by 7. Thus, 1050, 1155, 1260 ; and 3010, 4214, 4816.

9. When the sum of all the *odd* places in any number, is equal to the sum of all the *even* places, the whole is divisible by 11.

Thus, 879615. $5+6+7=1+9+8$.

2. Reduce $\frac{192}{576}$ to its lowest terms.   Ans. $\frac{1}{3}$.

3. Reduce $\frac{252}{364}$ to its lowest terms.   Ans. $\frac{9}{13}$.

4. Reduce $\frac{1344}{1536}$ to its lowest terms.   Ans. $\frac{7}{8}$.

It is obvious, that either multiplying or dividing both the numerator and denominator of a fraction by the same number, does not change its value. Thus, $\frac{1}{2}$ multiplied continually by 2 becomes $\frac{2}{4}$, $\frac{4}{8}$, $\frac{8}{16}$, $\frac{16}{32}$, $\frac{32}{64}$, $\frac{64}{128}$, and so on ; in which series, as the equal parts, designated by the denominators are successively lessened one half, the number taken, (designated by the numerator,) is correspondently increased.

Reversing the process, and dividing continually by 2, will of course, restore the original fraction ; $\frac{64}{128} \div 2 = \frac{32}{64}$, $\frac{16}{32}$, $\frac{8}{16}$, $\frac{4}{8}$, $\frac{2}{4}$, $\frac{1}{2}$.

II. *To reduce a whole number to a fraction of a specified denominator.*

RULE. Multiply the whole number by the given denominator, and take the product for a numerator.

1. Reduce 15 to a fraction whose denominator is 7.
$$15 \times 7 = 105.$$ Ans. $\frac{105}{7}$.

2. Reduce 8 to thirds. $8 \times 3 = 24.$ Ans. $\frac{24}{3}$.

3. Reduce 12 to fifths.   Ans. $\frac{60}{5}$.

4. How many eighths of a yard in 12 yards ?   Ans. $\frac{96}{8}$.

5. In 13 feet, how many $\frac{1}{12}$ of a foot?   Ans. $\frac{156}{12}$.

III. *To reduce a mixed number to an equivalent fraction.*

RULE. Multiply the whole number by the denominator of the fraction, add the numerator, and the sum will form the numerator of the fraction required.

1. What fraction is equivalent to $5\frac{4}{7}$?
$$5 \times 7 = 35 + 4 = 39.$$ Ans. $\frac{39}{7}$.

2. In $6\frac{3}{8}$, how many eighths ?   Ans. $\frac{51}{8}$.

3. In twenty bushels and three fourths of a bushel of wheat, how many fourths ?   Ans. $\frac{83}{4}$.

4. In $11\frac{5}{8}$ dollars, how many eighths ?   Ans. $\frac{93}{8}$.

5. Reduce $7\frac{3}{5}$ yards to fifths.   Ans. $\frac{38}{5}$.

IV. *To reduce an improper fraction to its equivalent whole or mixed number.*

RULE. Divide the numerator by the denominator.

1. How many units are there in $\frac{36}{9}$?
$$36 \div 9 = 4. \qquad\qquad \text{Ans. 4.}$$

2. How many bushels of wheat are there in $\frac{83}{4}$ of a bushel ? Ans. $20\frac{3}{4}$.

Since there are $\frac{4}{4}$ in 1 bushel, there will be $\frac{1}{4}$ as many whole ones as there are fourths.

3. In $\frac{72}{9}$ how many whole ones ? In $\frac{72}{8}$ how many ?

4. In $\frac{19}{5}$ of an ell, how many ells ? Ans. $3\frac{4}{5}$.

V. *To reduce a compound fraction to an equivalent simple one.*

RULE. Multiply the numerators together for the numerator, and the denominators together for the denominator of the required fraction.

When a whole or mixed number is connected with the question, it must first be reduced to an improper fraction.

Any two terms of the fraction may be divided by the same number, and the quotients be substituted for them; or, if there are two that are common they may be dropped.

1. Reduce $\frac{1}{2}$ of $\frac{2}{3}$ of $\frac{3}{4}$ to a simple fraction.

$$\frac{1\times2\times3}{2\times3\times4}=\frac{6}{24}=\frac{1}{4} \text{ Ans.} \qquad \text{Or,} \quad \frac{1\times2\times3}{2\times3\times4}=\frac{1}{4},$$

by dropping like terms above and below the line.

2. Reduce $\frac{1}{2}$ of $\frac{2}{3}$ of $\frac{3}{7}$ of $5\frac{5}{6}$ to a simple fraction.

$$\frac{1\times2\times3\times35}{2\times3\times7\times6}=\frac{5}{6} \text{ Ans.}$$

Here dropping the *twos* and the *threes* on each side of the line, and dividing 35 by 7, we have the answer in the lowest terms.

3. Find the simple expression for $\frac{3}{4}$ of $\frac{5}{9}$ of 8 dollars.
Ans. $\frac{10}{3}$, or $ 3\frac{1}{3}$.

In this example, dividing the 9 by 3, and the 8 by 4, the fraction will stand $\frac{5\times2}{3}=\frac{10}{3}$.

4. How much is $\frac{2}{3}$ of $\frac{5}{8}$ of $3\frac{1}{2}$ yards of cloth ?   Ans. $\frac{7}{8}$.

5. Reduce $\frac{5}{6}$ of $\frac{3}{20}$ of $42\frac{2}{3}$ to a simple fraction.

Ans. $\frac{53}{10} = 5\frac{3}{10}$.

VI. *To find the value of a fraction in parts of the integer.*

RULE. Multiply the numerator by the parts in the next inferior denomination, and divide the product by the denominator; and multiply the remainder—if any—by the parts in the *next* denomination, and so on. The quotients placed in order, will express the value of the fraction.

1. What is the value of $\frac{2}{3}$ of £1.*

By *integer* is meant unity of that, of which the fraction expresses a part. In this example it is one pound, the parts of which are 20 shillings.

$$
\begin{array}{r}
2 \\
20 \\
\hline
3)40(13s.\ 4d.\ \text{Ans.} \\
39 \\
\hline
1 \\
12 \\
\hline
3)12(4d. \\
12 \\
\hline
\end{array}
$$

The remainder (1) after the first division, is $\frac{1}{3}$, forming a new fraction, of which the integer is 1 shilling.

2. What is the value of $\frac{2}{3}$ of a guinea ?   Ans. 4s. 8d.

3. What is the value of $\frac{3}{5}$ of 1 lb. Troy ?

Ans. 7 oz. 4 dwt.

4. In $\frac{7}{16}$ of 1 cwt. how many quarters, pounds, &c. ?

Ans. 1 qr. 21 lb.

5. How much is $\frac{6}{7}$ of a hogshead ?   Ans. 54 gal.

Here, instead of multiplying 6 by 63, and dividing by 7. divide 63 by 7, and multiply the product by 6, according to the method of contraction given in the Supplement to Division.

---

* Observe that $\frac{2}{3}$ of £1 = $\frac{40}{3}$ (that is $\frac{2 \times 20}{3}$) of 1s., because a pound is 20 times the value of a shilling.

6. What is the value of $\frac{13}{16}$ of a cord?   Ans. 104 ft.

$$\text{Here,} \quad \frac{13 \times 128}{16} = 13 \times 8 = 104.$$

7. How many quarters and nails in $\frac{7}{10}$ of an ell English?
                        Ans. 3 qr. 2 n.

8. How many shillings and pence, New York currency, are $\frac{5}{16}$ of a dollar?   Ans. 2s. 6d.

VII. *To reduce lower denominations to the fraction of a higher denomination.*

RULE. Reduce the given quantity to the lowest denomination mentioned, for the numerator; then reduce the integer to the same denomination, for the denominator of the required fraction.

1. What part of a pound is 3s. 4d.?

$$3s. \ 4d. \quad = 40d.$$
$$\text{Integer} \quad £1 \times 20 \times 12 = 240d.$$

$$\frac{\text{Numerator}}{\text{Denominator}} \quad \frac{40}{240} = \frac{1}{6} \text{ Ans.}$$

2. What part of a yard is 3 qr. 2$\frac{1}{2}$ na.?
$$3 \text{ qr. } 2\frac{1}{2} \text{ na.} = 29 \text{ half nails}$$
$$1 \text{ yd.} = 32 \text{ half nails.} \quad\quad\quad \text{Ans. } \frac{29}{32}.$$

3. What part of 1 lb. Troy, is 7 oz. 4 dwt.?   Ans. $\frac{3}{5}$.
4. Reduce 3 qr. 7 lb. to the fraction of 1 cwt.
                            Ans. $\frac{91}{112}$.
5. What part of 1 rod is 4 yd. 1$\frac{1}{2}$ ft?   Ans. $\frac{9}{11}$.
6. What part of 1 bushel is 1$\frac{3}{5}$ pk.?   Ans. $\frac{2}{5}$.
7. Reduce 13 cwt. 3 qr. 20 lb. to the fraction of a ton.
                            Ans. $\frac{35}{56}$.
8. Reduce 1 hhd. 49 gal. of wine to the fraction of a tun.
                            Ans. $\frac{4}{5}$.
9. What part of 4 cwt. 1 qr. 24 lb. is 3 cwt. 3 qrs. 17 lb 8 oz.?
                            Ans. $\frac{7}{8}$.

VIII. *To reduce a fraction from one denomination to another.*

RULE. By the tabular parts between the given quantity and the denomination to which it is to be reduced, multiply either the numerator or the denominator, according as the reduction is from higher to lower, or from lower to higher denominations.

1. Reduce $\frac{2}{9}$ of a pound to the fraction of a penny.

$$\frac{2 \times 20 \times 12}{9} = \frac{480}{9} = \frac{160}{3} \text{ Ans.}$$

2. Reduce $\frac{2}{9}$ of a penny to the fraction of a pound.

$$\frac{2}{9 \times 12 \times 20} = \frac{2}{2160} = \frac{1}{1080} \text{ Ans.}$$

In the first example, the reduction is from a higher to a lower denomination; therefore the numerator is multiplied. In the second, the reduction is from lower to higher, therefore, the denominator is multiplied.

The principle of the rule will be understood, if we consider that $\frac{2}{9}$ of a pound must be 20 times as many ninths of a shilling, and so on. But, $\frac{2}{9}$ of a *penny* can be but $\frac{1}{12}$ as many ninths of a shilling, and the parts of a shilling, but $\frac{1}{20}$ as many parts of a pound.

3. What fraction of 1 lb. is $\frac{2}{7}$ of 1 cwt.

$$\frac{2 \times 4 \times 28}{7} = \frac{32}{1} \text{ Ans.}$$

Observe, that 28 is divisible by 7, and by substituting the quotients, you get the answer in the *lowest terms*, and with the least trouble.

4. Reduce $\frac{4}{5}$ dwt. to the fraction of 1 lb.   Ans. $\frac{1}{300}$.
5. Reduce $\frac{4}{5}$ *l.* to the fraction of a guinea.

$$\frac{4 \times 20}{5 \times 21} = \frac{16}{21} \text{ Ans.}$$

6. What part of a yard, is $\frac{4}{7}$ of an inch?   Ans. $\frac{1}{63}$.
7. Reduce $\frac{5}{8}$ of a crown to the fraction of a guinea.
                                                    Ans. $\frac{25}{168}$.
9 What part of a bushel is $\frac{2}{3}$ of a quart?   Ans. $\frac{1}{48}$.

IX. *To reduce fractions to a common denominator.*

RULE. Multiply each numerator into all the denominators except its own for the new numerators; and multiply
all the denominators together for a common denominator.

1. Reduce $\frac{2}{3}$, $\frac{1}{2}$, and $\frac{3}{4}$ to a common denominator.

$$2\times2\times4=16 \text{ the numerator for } \tfrac{2}{3}$$
$$1\times3\times4=12 \quad - \quad - \quad \text{do.} \quad - \quad - \quad \tfrac{1}{2}$$
$$3\times3\times2=18 \quad - \quad - \quad \text{do.} \quad - \quad - \quad \tfrac{3}{4}$$
$$3\times2\times4=24 \text{ the common denominator.}$$

The equivalent fractions are $\begin{cases} \frac{16}{24}=\frac{2}{3} \\ \frac{12}{24}=\frac{1}{2} \\ \frac{18}{24}=\frac{3}{4} \end{cases}$

It is evident in this case, that the value of the fractions
is not changed, because both the terms of each, are multiplied *by the same numbers.*

2. Reduce $\frac{2}{7}$ and $\frac{4}{9}$ to a common denominator.

Ans. $\begin{cases} \frac{18}{63}=\frac{2}{7} \\ \frac{28}{63}=\frac{4}{9} \end{cases}$

3. Reduce to a common denominator $\frac{5}{6}$, $\frac{1}{3}$, and $\frac{1}{2}$.

Ans. $\frac{5}{6}$, $\frac{2}{6}$, $\frac{3}{6}$.

Here, the largest denominator 6, may be divided by each
of the others; consequently, the two latter fractions may
be reduced to sixths; the first requiring no reduction.

4. Reduce $\frac{3}{8}$, $\frac{1}{4}$, $\frac{5}{6}$, and $\frac{5}{12}$.　　　Ans. $\frac{9}{24}$, $\frac{6}{24}$, $\frac{20}{24}$, $\frac{10}{24}$.

5. Reduce $2\frac{2}{5}$ and $4\frac{1}{3}$ to a common denominator.

Ans. $\frac{36}{15}$, $\frac{65}{15}$.

6. Reduce $\frac{2}{5}$ of $\frac{3}{5}$ and $\frac{1}{4}$ of 3.　　　Ans. $\frac{8}{20}$, $\frac{15}{20}$.

ADDITION OF VULGAR FRACTIONS.

BEFORE fractions can be added, they must be of the same
denomination, that is *like* parts of the same integer. We
cannot, for example, add together 1 half and 2 thirds; because they would make neither 3 halves, nor 3 thirds. It
is necessary, therefore, to bring them to a common denominator.

Before entering upon this branch, it will be useful to understand the following Problem : *To find the least common multiple of two or more numbers ; that is, the least number divisible by them without remainder.*

RULE. Place the numbers in the order of their values the highest first ; compare each of the lower numbers with any one above it, and if it will measure the higher exactly, it may be omitted ; if not, divide each of the inferior numbers by the greatest common measure of it and any superior number ; and set the quotients in a line beneath, together with any number that is prime* to all above it ; then the continued product of the numbers beneath, and the highest number, will be the least common multiple.

1. What is the least common multiple of 4, 5, 6, and 10 ?

$$10 \quad 6 \quad 5 \quad 4$$
$$\overline{10 \times 3 \quad \times \quad 2 = 60 \text{ Ans.}}$$

2. Find the least common multiple of 12, 25, 30, and 45 ?

$$45 \quad 30 \quad 25 \quad 12$$
$$\overline{45 \times 2 \times 5 \times 2 = 900 \text{ Ans.}}$$

In this example, 15 measures 30 and 45 ; 5 measures 25 and either of the numbers above ; and 6 measures 12 and 30.

3. Find the least common multiple of 3, 5, 8, and 12.
Ans. 120.

4. What is the least common multiple of 12, 6, 4, and 3 ?
Ans. 12.

This problem is of use in ascertaining the *least common denominator* of several fractions ; which is in all cases equal to the least common multiple of the denominators. Divid-

---

* Numbers are said to be prime to one another, when unity is the only common number that will measure them.

ing this by the denominators severally, and multiplying the quotient by the numerator of the respective fractions, we have also the required numerators.

5. Reduce $\frac{3}{4}$, $\frac{5}{8}$, $\frac{2}{7}$, and $\frac{8}{21}$ to equivalent fractions having the least common denominator.

$$\overset{2}{\underline{1\ 8\ 7\ 4}}$$
$$21\times8=\ 168\ \text{the least common denominator,}$$

$$168\div\begin{cases} 4\times3=126 \\ 8\times5=105 \\ 7\times2=\ 48 \\ 21\times8=\ 64 \end{cases} \qquad \text{Ans.}\begin{cases} \frac{126}{168}=\frac{3}{4} \\ \frac{105}{168}=\frac{5}{8} \\ \frac{48}{168}=\frac{2}{7} \\ \frac{64}{168}=\frac{8}{21} \end{cases}$$

To add vulgar fractions.

RULE. Reduce compound fractions to simple ones, and all the fractions to their least common denominator; then the sum of the numerators placed above the common denominator, will be the sum of the fractions required.

1. Add $\frac{1}{2}$, $\frac{2}{3}$, and $\frac{3}{4}$ together.

$$\frac{1}{2}=\frac{6}{12} \quad \frac{2}{3}=\frac{8}{12} \quad \frac{3}{4}=\frac{9}{12}.$$

$$\frac{6}{12}+\frac{8}{12}+\frac{9}{12}=\frac{23}{12}=1\frac{11}{12} \text{ Ans.}$$

2. Add together $\frac{3}{4}$, $\frac{5}{8}$, $\frac{2}{7}$, and $\frac{8}{21}$.
See example 5, under the problem.

$$\frac{126}{168}+\frac{105}{168}+\frac{48}{168}+\frac{64}{168}=\frac{343}{168}=2\frac{1}{24} \text{ Ans.}$$

3. Add $\frac{3}{5}+\frac{5}{6}+\frac{2}{3}$.   Ans. $\frac{63}{30}=2\frac{1}{10}$.
4. Add $\frac{7}{12}+\frac{11}{18}+\frac{19}{36}$.   Ans. $12\frac{11}{18}$.
5. Add $3\frac{2}{5}$ yd. $1\frac{7}{8}$ yd. and $\frac{9}{10}$ yd. together.

NOTE. Find the sum of the fractions separately from the integers.

$$\frac{2}{5}+\frac{7}{8}+\frac{9}{10}=2\frac{7}{40}+3+1=6\frac{7}{40} \text{ yd. Ans.}$$

6. Add $6\frac{1}{2}+5\frac{1}{3}+1\frac{1}{2}$.   Ans. $12\frac{17}{18}$.
7. What is the sum of $\frac{2}{3}$ of a pound and $\frac{5}{6}$ of a shilling?
     Ans. $13\frac{5}{6}$s. or 13s. 10d. $2\frac{2}{3}$q.

It is evident that these fractions must be reduced to the same denomination; because *parts* of a pound, and *parts* of a shilling cannot be put together, any more than pounds and shillings can be united in one sum.

8. Add $\frac{2}{5}$ of a mile, $\frac{1}{4}$ of a furlong, and $\frac{1}{2}$ of a rod.

Ans. $138\frac{1}{2}$ rods.

9. What is the sum of $\frac{7}{8}$ of a yard, and $\frac{3}{8}$ of a quarter.

Ans. $1\frac{1}{24}$ yd.

SUBTRACTION OF VULGAR FRACTIONS.

Rule. Prepare the fractions as for addition; and the difference of the numerators placed over the common denominator, will express the difference of the fractions.

1. From $\frac{3}{4}$ take $\frac{5}{7}$.

$$\frac{3}{4}=\frac{21}{28}, \quad \frac{5}{7}=\frac{20}{28} \quad \frac{21}{28}-\frac{20}{28}=\frac{1}{28} \text{ Ans.}$$

2. From $\frac{5}{6}$ take $\frac{7}{12}$. Ans. $\frac{3}{12}$ or $\frac{1}{4}$.
3. From $\frac{13}{30}$ take $\frac{2}{5}$ of $\frac{7}{12}$. Ans. $\frac{1}{5}$.
4. From $5\frac{3}{8}$ yards take $3\frac{3}{4}$ yards.

As the fraction in the minuend, is less than the fraction in the subtrahend, we may take a unit from the integer of the former, and add its value to the fraction. $1=\frac{8}{8}+\frac{3}{8}=\frac{11}{8}$. Then $4\frac{11}{8}-3\frac{3}{4}=1\frac{5}{8}$ Ans.

5. From £$9\frac{1}{2}$ take £$6\frac{9}{10}$. Ans. £$2\frac{3}{10}$.
6. What is the difference between $\frac{3}{4}$ of $5\frac{1}{3}$ and $\frac{2}{5}$ of $8\frac{1}{8}$?

Ans. $1\frac{4}{5}$.

7. From $\frac{5}{6}$ of a pound, take $\frac{2}{5}$ of $\frac{3}{4}$ of a shilling.

Ans. £$\frac{191}{360}$=10s. 7d. $1\frac{1}{3}$qr.

8. From $\frac{5}{6}$ of a yard, take $\frac{3}{4}$ of a quarter.

Ans. $\frac{14}{8}$=1 qr. 3 na.

Both in addition and subtraction, the value of the fractions may be found separately in parts of the integer. and these be added or subtracted as compound numbers.

9. From $\frac{3}{8}$ of a pound, take $\frac{5}{6}$ of a shilling.

£$\frac{3}{8}$=7s. 6d. $\frac{5}{6}$ 1s.=10d.   Ans. 6s. 8d.

10. From $1\frac{3}{4}$qr. take $\frac{5}{16}$ of 1 cwt.   Ans. $\frac{1}{2}$qr.=14 lb.

(6)

MULTIPLICATION OF VULGAR FRACTIONS.

IT has already been shown, that multiplying or dividing, both the terms of a fraction, by the same number, does not change its value. But multiplying or dividing *one* of the terms, the other remaining the same, changes the value of a fraction.

If the numerator be increased, the value of the fraction is increased: if the denominator be increased the value of the fraction is diminished. Thus, the fraction $\frac{2 \times 2}{6} = \frac{4}{6} = \frac{2}{3}$ and $\frac{2}{6 \times 2} = \frac{2}{12} = \frac{1}{6}.$ The reason is plain; multiplying the numerator, increases the number of parts taken, the parts themselves remaining undiminished; and multiplying the denominator, lessens the quantity of the equal parts, while only the same number is taken.

Division has just the reverse effect:—thus, $\frac{2 \div 2}{6} = \frac{1}{6}$ and $\frac{2}{6 \div 2} = \frac{2}{3}.$ Hence, it is evident, that multiplying the numerator, or dividing the denominator of a fraction, has each the effect to *increase* its value equally; and dividing the numerator or multiplying the denominator, the effect to *diminish* its value equally. Therefore—

I. *To multiply a fraction by a whole number.*

RULE. Multiply the numerator, or divide* the denominator.

1. Multiply $\frac{1}{8}$ by 4.

$$\frac{1 \times 4}{8} \quad \frac{4}{8} = \frac{1}{2} \text{ and' } \frac{1}{8 \div 4} = \frac{1}{2} \text{ Ans.}$$

The second method presents the answer in the *lowest terms* without reduction.

---

* This can only be done, when the denominator is an exact *multiple* of the multiplier.

2. At $\frac{7}{8}$ of a dollar a yard, what will 5 yards of linen cost ?   Ans. $\frac{35}{8}$, or $ 4$\frac{3}{8}$.

As in multiplication, it is indifferent, which of the two given factors we make the multiplier; it follows, that a whole number is multiplied by a fraction in the same manner as a fraction by a whole number.

3. Multiply 4 by $\frac{1}{8}$.   Ans. $\frac{1}{2}$.

4. At $ 5 a yard, what will $\frac{7}{8}$ of a yard come to ?

Ans. $ 4$\frac{3}{8}$.

5. Multiply 35 by $\frac{3}{5}$.   Ans. 21.

II. *To multiply one fraction by another.*

Rule. Multiply the numerators together for the numerator of the required fraction, and the denominators together for the denominator.

1. Multiply $\frac{3}{4}$ by $\frac{2}{5}$.   Ans. $\frac{6}{20} = \frac{3}{10}$.

Here $\dfrac{3 \times 2}{4} = \dfrac{6}{4}$. But the multiplier is only the $\frac{1}{5}$ part of 2; consequently $\frac{6}{4}$ must be diminished correspondently, by multiplying its denominator by 5.

2. Multiply $2\frac{1}{2}$ by $\frac{3}{4}$ of $\frac{2}{5}$ of 8.

$$\frac{5 \times 3 \times 2 \times 8}{2 \times 4 \times 5 \times 1} = \frac{3 \times 2}{1} = 6 \text{ Ans.}$$

The mixed number $2\frac{1}{2}$ reduced to an improper fraction is $\frac{5}{2}$, and the integer 8 is made a fraction, by putting 1 for its denominator.

Cancelling like terms and dividing 8 and 4 by 4, we have the result in the lowest terms.

3. Multiply $\frac{9}{11}$ by $\frac{3}{4}$ of $7\frac{1}{3}$.   Ans. $\frac{9}{2}$ or $4\frac{1}{2}$.

4. Multiply $4\frac{9}{10}$ by $6\frac{4}{7}$.   Ans. $32\frac{1}{5}$.

5. If a horseman travel at the rate of $7\frac{2}{3}$ miles an hour, for $6\frac{7}{12}$ hours; how far will he have gone?   Ans. $50\frac{1}{36}$.

6. What is the cost of $9\frac{1}{8}$ yards of linen, at $ 1$\frac{3}{8}$ per yd. ?

Ans. $ 12$\frac{35}{64}$.

7. Required the product of 6, and $\frac{2}{3}$ of 5.   Ans. 20

8. Required the product of $\frac{7}{9}$, $\frac{3}{8}$ and $4\frac{5}{14}$.   Ans. $2\frac{1}{30}$.

9. Multiply together 5, $\frac{2}{3}$, $\frac{6}{7}$ of $\frac{3}{5}$, and $4\frac{1}{8}$.

Ans. $\frac{50}{7} = 7\frac{1}{7}$.

DIVISION OF VULGAR FRACTIONS.

I. *To divide a fraction by a whole number.*

RULE. Multiply the denominator by the given number; or,—when it can be done,—divide the numerator.

1. If $\frac{6}{7}$ of a barrel of flour be divided equally between two poor families ; what will be the share of each ?

$$\frac{6}{7\times2}=\frac{6}{14}=\frac{3}{7}: \quad \text{Or,} \quad \frac{6\div2}{7}=\frac{3}{7} \text{ Ans.}$$

2. Divide $\frac{15}{6}$ by 5.   Ans. $\frac{3}{16}$.
3. Divide $\frac{15}{16}$ by 4.   Ans. $\frac{15}{64}$.

II. *To divide a whole number by a fraction.*

RULE. Multiply the dividend by the denominator of the fraction, and divide the product by the numerator.

1. Divide 8 by $\frac{3}{4}$.

$$\frac{8\times4}{3}=\frac{32}{3} \text{ Ans.}$$

To understand the principle of this operation :—had the divisor been $\frac{1}{4}$, the question would simply have been, how many $\frac{1}{4}$ there are in 8. Obviously, there are 4 times as many as there are units in the number. But there can be only one third as many $\frac{3}{4}$ in any number, as there are $\frac{1}{4}$; consequently the numerator is placed under the product.

2. If you have $ 6 to purchase oats at $ $\frac{3}{8}$ per bushel; how many bushels can you buy ?

$$\frac{6\times8}{3}=16$$

            Ans. 16 bushels.

3. Divide 12 by $\frac{2}{3}$ Ans. $\frac{36}{2}=18$.
4. Divide 16 by $\frac{6}{7}$ Ans. $18\frac{2}{3}$.

III. *To divide one fraction by another.*

RULE. Multiply the numerator of the dividend into the denominator of the divisor for the numerator of the required fraction ; and the denominator of the dividend into the numerator of the divisor for the required denominator.

For convenience, place the numerator of the dividend above a horizontal line, and its denominator below, and place the divisor in the reverse order: then cancel and reduce the terms as far as practicable, before multiplication.

1. Divide $\frac{14}{9}$ by $\frac{7}{15}$.

$$\frac{14 \times 15}{9 \times 7} = \frac{2 \times 5}{3 \times 1} = \frac{10}{3} \text{ or, } 3\frac{1}{3} \text{ Ans.}$$

2. Divide $8\frac{2}{3}$ by $6\frac{1}{2}$.

$$8\frac{2}{3} = \frac{26}{3}$$
$$6\frac{1}{2} = \frac{13}{2}$$

$$\frac{26 \times 2}{3 \times 13} = \frac{2 \times 2}{3} = 1\frac{1}{3} \text{ Ans.}$$

To explain the principle of this rule:—Take as an example $\frac{2}{5} \div \frac{3}{4}$. Now if $\frac{2}{5}$ were divided by 3, considered as a whole number, the result would be $\frac{2}{15}$: but the divisor is only the $\frac{1}{4}$ part of 3 ; consequently the quotient in this case, should be 4 times as large as in the other, and it is made so by multiplying the numerator by 4, the denominator of the divisor. $\frac{8}{15}$ Ans.

3. Divide $\frac{3}{7}$ of $\frac{1}{2}$ of 5, by $2\frac{1}{4}$.   Ans. $\frac{10}{21}$.
4. If a man travel at the rate of $7\frac{2}{5}$ miles an hour, in how many hours will he have gone $50\frac{1}{30}$ miles ?
              Ans. $6\frac{7}{12}$ hours.
5. How many times is $5\frac{1}{2}$ contained in $53\frac{1}{3}$?
              Ans. $9\frac{23}{33}$.
6. How many times is $\frac{2}{15}$ contained in $\frac{14}{15}$?
              Ans. 7 times.
7. Divide $\frac{25}{27}$ by $\frac{5}{9}$.   Ans. $\frac{5}{3} = 1\frac{2}{3}$.

NOTE. In this example, each of the terms of the dividend is a multiple of the corresponding term of the divisor. In such cases, the terms may be divided directly.

8. What quantity of wheat at $\frac{7}{8}$ of a dollar a bushel, can be bought with $ $11\frac{3}{4}$?   Ans. $13\frac{3}{7}$ bu.
9. How many times is $\frac{2}{3}$ of a pint contained in $\frac{5}{9}$ of a gallon ?   Ans. $6\frac{2}{3}$.

The fractions must be brought to the same denomination before division ; that is, both must be parts of a gallon, or parts of a pint.

(6*)

10. How many times can a vessel holding $\frac{9}{10}$ of a quart, be filled from $\frac{1}{5}$ of a barrel, containing $31\frac{1}{2}$ gallons ?

Ans. $46\frac{2}{3}$.

11. How many rods are there in $49\frac{1}{2}$ yards ?. Ans. 9.

QUESTIONS.

What are fractions ?
What does the denominator of a fraction denote ?
What the numerator ?
If both the terms of a fraction are multiplied by the same number, is its value changed ?
If both are divided by the same number, what is the effect ?
If the numerator *only* be increased, what is the effect ?
What, if the denominator only be increased ?
Which would you prefer $\frac{9}{10}$, or $\frac{9}{20}$ of a prize in a lottery ?
What are the methods of multiplying a fraction by a whole number ?
Which of them is to be preferred ?
How is a whole number multiplied by a fraction ?
How is a fraction divided by a whole number ?
How is a whole number divided by a fraction ?
Can fractions of different integers—for example, parts of a yard, and parts of a foot—be added or subtracted *immediately* ?
Can fractions with different denominators be thus added or subtracted ?
What is the rule for finding the common denominator of several fractions ?

DECIMAL FRACTIONS.

A DECIMAL FRACTION is one, whose denominator is 10, 100, 1000, &c. Its denominator is not written, but a point is prefixed to the numerator to designate the decimal character of its denominator. Thus, $\frac{5}{10}$, $\frac{25}{100}$, $\frac{164}{1000}$, are written .5, .25, .164.

When the numerator has not so many places of figures, as the denominator has ciphers, the deficiency is supplied by prefixing ciphers : thus, $\frac{5}{100}$, $\frac{5}{1000}$, are written .05, .005. It will be perceived, that every additional cipher removes the numerator one place farther from the decimal point, and diminishes the value of the fraction tenfold.

In a mixed number, as $6\frac{4}{10}$, the decimal point is placed between the integer and the fraction; thus, 6.4 : hence it is sometimes called *separatrix*.

NUMERATION TABLE.

Millions	C Thousands	X Thousands	Thousands	Hundreds	Tens	Units	Tenth parts	Hundredth parts	Thousandth parts	X Thousandth parts	C Thousandth parts	Millionth parts
7	6	5	4	3	2	1 .	2	3	4	5	6	7

Whole numbers.   Decimals.

The decimal system of notation, according to which the value of figures decreases from left to right, tenfold for each remove towards the unit's place, was explained under the head of numeration. It will be seen from the table above, that the same system is continued below the unit to express the decimal divisions of *it*, and of the successive divisions themselves. To understand these decimal fractions, we have only to suppose that 1 (apple, bushel, or any thing else) is divided into 10 equal parts, each of these parts into ten other equal parts, and each of these last into 10 more, and so on: then the figure next below the unit's place, will denote parts of the first division, that is, tenths; the next figure towards the right will denote divisions of the tenths, that is, hundredths; the third place of decimals, will denote divisions of the hundredths, and so on; the equal parts continually decreasing tenfold for every additional remove from the unit's place.

The reason, therefore, is plain, why such divisions of a unit are called *Decimal* fractions, that word signifying belonging to *ten*.

Taking 1 as the starting point in numeration, we see that the order of increase, and of decrease on the left and right of it, is correspondent. On the left we have tens, on the right, *tenths*; hundreds, *hundredths*; thousands, *thousandths*; and so on at equal distances from the unit's place.

Decimals may be read separately by their places; but it

is usual to read them precisely as whole numbers, adding at the end, the name of the denominator of the *last figure.* Thus .23, is twenty-three hundredths, instead of two tenths three hundredths. The two expressions are exactly equivalent, because $\frac{2}{10} = \frac{20}{100}, \frac{200}{1000}$, and so on; as will be evident from reducing the fractions to their *lowest terms.*

Ciphers placed at the right hand of a decimal fraction do not alter its value, since every significant figure continues to possess the same place: so .5 .50 and .500 are all of the same value, and equal to $\frac{5}{10}$ or $\frac{1}{2}$.

NOTE. It may assist the learner in reading decimals, to write them first as vulgar fractions; making the denominator consist of 1, with as many ciphers as there are places *after* the decimal point.

Write the following numbers as decimals.

$$\frac{16}{100}, \frac{16}{1000}, \frac{16}{10000}, \frac{16}{1000000}, \frac{3}{10}, \frac{10}{1000}.$$
$$19\frac{4}{10}, 19\frac{4}{100}, 19\frac{44}{1000}.$$

Read the following numbers.

2 8 4 5.6	2.8 4
2 8 4.5 6	.2 8 4
2 8.4 5 6	.0 2 8 4 5 6

ADDITION OF DECIMALS.

RULE. Place the numbers, whether mixed or pure decimals, so that their decimal points may fall under one another, add as in whole numbers, and place the decimal point in the sum, under the decimal points above.

1. Add together 807.2659+70.602+4.06+151.7

```
      8 0 7.2 6 5 9        (2)  .1 9 9
       7 0.6 0 2                2.7 5 6 9
          4.0 6                  .2 5
      1 5 1.7                    .6 5 4
```

Ans. 1 0 3 3.6 2 7 9   Ans. 3.8 5 9 9

3. What is the sum of 376 .25+86 .125+637 .4725+ 6 .5+41 .02+358 .865 ?   Ans. 1506 .2325.

4. Add $ 3 .5+$ 47 .25+$ 2 .0073+$ 927 .01+$ 1 .5.
        Ans. $ 981 .2673.

5. Required the sum of .0463+12 .5+.65+112+54 .321+276.   Ans. 455 .5173.

SUBTRACTION OF DECIMALS.

RULE. Place the numbers as in addition of decimals, subtract as in whole numbers, and place the decimal point in the remainder under the decimal points above.

1. From 91 .73 take 2 .138.

        9 1.7 3
        2.1 3 8
        ——————
Ans. 8 9.5 9 2

(2) 2.7 3
        1.9 1 8 5
        ——————
Ans. .8 1 1 5

3. Find the difference between 714 . and .916.
        Ans. 713 .084.

4. From .145 take .09684.   Ans. .04816.

5. How much greater is 2 than .298 ?   Ans. 1 .702.

MULTIPLICATION OF DECIMALS.

RULE. Whether they be mixed numbers, or pure decimals, place the factors and multiply them as in whole numbers, and point off so many figures from the product as there are decimal places in both the factors ; and if there be not so many places in the product, supply the defect by prefixing ciphers to the left hand.

1. Multiply 3 .024
        by 2 .23
        ——————
Ans. 6 .74352

2. Multiply 5 .236
        by .008
        ——————
Ans. .041888

The reason of pointing off decimals for the decimal places in the multiplicand, is sufficiently evident. But, we are directed to point off likewise for the decimals in the multiplier. Taking the second example, you will observe, that had the multiplier (8) been *units*, there should have been 3 decimal places in the product; but as the 8 stands in the third place of decimals, it is only 8 *thousandths*, the product, therefore, should be as much less than in the other case, as thousandths are less than units. Throwing the decimal point 1 place toward the left diminishes the value of every figure in the product tenfold, and putting it 3 places toward the left diminishes it a thousandfold, which it evidently should be, because the multiplier is only thousandths of a unit. But, in order to remove the separatrix the requisite number of places to the left, it is necessary to prefix the cipher.

3. Multiply 79 .347 by 23 .15.   Ans. 1836 .88305.
4. Multiply $ 341 .45 by .007.   Ans. $ 2 .39015.
5. Multiply .385746 by .00464.

Ans. .00178986144.

6. At 7 cents a dozen, what is the cost of 26 .5 dozen of eggs ?   Ans. $ 1 .855.

7. At $5 .47 a barrel, what is the cost of 83 barrels of flour ?   Ans. $ 454 .01.

8. Multiply .09 by .7.   Ans. .063.

9. At $4\frac{1}{2}$ mills a piece, what will 350 quills cost ?

Ans. $ 1 .57$\frac{1}{2}$.

To multiply by 10, 100, 1000, &c.

RULE. Remove the decimal point as many places toward the right, as there are ciphers in the multiplier.

NOTE. Removing the separatrix one place toward the right, increases the value of each figure tenfold, or multiplies it by 10 ; removing it two places, increases the value one hundredfold, and so on.

$$\text{Thus,} \ .365\times \begin{cases} 10 & =3.65 \\ 100 & =36.5 \\ 1000 & =365. \\ 10000 & =3650. \end{cases}$$

DIVISION OF DECIMALS.

RULE. Divide as in whole numbers ; and point off from the right of the quotient as many places for decimals, as the decimal places in the dividend exceed those in the divisor.

If there are not as many places in the quotient as the rule requires, prefix ciphers.

When there are more decimals in the divisor, than in the dividend, annex ciphers to the latter to make its decimal places equal to those of the divisor ; and ciphers may also be annexed to the remainder—if any—and the division continued. The ciphers so annexed are to be counted as decimals.

Division of decimals is precisely like division of whole numbers, except the placing of the separatrix. As this is a point, which gives not a little perplexity to learners, we shall endeavor to make it plain.

If 3 be divided by 3 the quotient is 1. If 3 be divided by .3, it is evident, that the quotient ought to be as much larger than 1. as .3 is less than 3. But, .3 is only the tenth part of 3 ; therefore, the quotient should be 10 instead of 1.

Hence, we are directed to make the decimal places in the dividend *at least* equal to those in the divisor. Therefore in the example above, we add a cipher to the dividend, which makes it in fact, tenths with the divisor,

$$3 .0 \div .3 = 10.$$

Again, if we make .3 the dividend, and 3 the divisor, it is evident, that it is not contained in .3, 1 time, but only a fractional part of 1 ; and that, since the dividend is only the tenth part of 3, (which gives a quotient of 1,) the quotient should only be the tenth part of the former quotient 1, that is, .1

Again, if we make the dividend .03, and divide by 3, the quotient ought to be only the hundredth part of 1, because the dividend is only the hundredth part of 3. Therefore, we prefix a cipher to it, thus, .01.

The effect of the different positions of the separatrix may be exhibited as follows :

$$321 \div 321 = 1$$

$321 \div 32.1 = 10$	$32.1 \div 321 = .1$
$321 \div 3.21 = 100$	$3.21 \div 321 = .01$
$321 \div .321 = 1000$	$.321 \div 321 = .001$

From these divisions it is evident, that where the decimal places in the divisor and dividend are equal, the quotient is whole numbers. When the places in the divisor exceed those in the dividend, the quotient has to be multiplied by 10, 100, &c. Or (which has the same effect) ciphers are added to the dividend, to make its decimal places equal to those in the divisor.

When the decimal places in the dividend exceed those in the divisor by 1, the quotient has to be divided by 10; when by two places, it has to be divided by 100, and so on.

1. Divide 2 .39015 by .007.   (2)

.007)2.390 15   26.5)1.855(.07 Ans.
                                      1 855
Ans.   341.45

In the second example, the decimals in the dividend exceed those in the divisor by 2 places; there must, therefore, be 2 decimals in the quotient, and a cipher is prefixed to make the requisite number of places.

3. Divide 1836.88305 by 79.347.   Ans. 23.15.
4. Divide .178986144 by 3.85746.   Ans. .0464.
5. Divide 274.855 by .7853.   Ans. 350.
6. Divide 11 by .55.   Ans. 20.
7. Divide .55 by 11.   Ans. .05.
8. Divide 9 by 450.

$$9.00 \div 450 = .02 \text{ Ans.}$$

9. If a contribution amounting to $ 36.72 be made by a congregation, consisting of 918 persons; how much is it a piece?   Ans. .04, or 4 cents.
10. If 275 lemons, cost $ 2.475; how much is it a piece?
                                         Ans. 9 mills.

To divide by 10, 100, 1000, *&c.*

RULE. Remove the decimal point in the dividend, so many places toward the left hand, as there are ciphers in the divisor.

$$365 \div \begin{cases} 10 & = 3\,6.5 \\ 100 & = 3.6\,5 \\ 1000 & =.3\,6\,5 \end{cases} \qquad 25 \div \begin{cases} 10 & = \\ 100 & = \\ 1000 & = \end{cases}$$

REDUCTION OF DECIMALS.

I. *To reduce a vulgar fraction to its equivalent decimal.*

RULE. Place a decimal point at the right of the numerator, annex ciphers and divide by the denominator.

1. Reduce $\frac{1}{8}$ to a decimal.

$$8)1.000$$

Ans. .125

NOTE. The value of any fraction is found by dividing the numerator by the denominator: thus, $\frac{8}{4}$ equals 2. But in a proper fraction, we cannot divide the numerator without first reducing it. Adding one cipher makes it tenths two ciphers, hundredths, and so on. The quotient, therefore, will be decimal parts.

2. Reduce $\frac{1}{4}$, $\frac{1}{2}$, and $\frac{3}{4}$, to equivalent decimals.

Ans. .25, .5, .75.

3. What decimal is equivalent to $\frac{5}{8}$? Ans. .625.

4. What decimal is equivalent to $\frac{3}{5}$ of $\frac{2}{3}$? Ans. .4.

NOTE. Reduce it to a simple fraction before division.

5. Reduce $\frac{5}{6}$ to a decimal expression.

$$6)5.0000$$

.8333*

Ans. .8333+.

---

* The remainder in this example being constantly the same, there is a constant recurrence of the same quotient figure. In some cases two or more figures recur alternately; as .060606. Such decimals are called *repeating* or *circulating decimals.* A complete division in these cases, can only be *approximated.*

(7)

6. Reduce $\frac{2}{11}$ to a decimal.   Ans. .181818+.

II. *To reduce a decimal to a vulgar fraction.*

RULE. Write the denominator under the decimal, and (disregarding the decimal point) reduce the fraction to its lowest terms.

1. What fraction is equivalent to .25 ?

$$\frac{25}{100}=\frac{1}{4} \text{ Ans.}$$

2. What fraction is equivalent to .85 ?   Ans. $\frac{17}{20}$.
3. Reduce .09375 to a vulgar fraction.   Ans. $\frac{3}{32}$.
4. Reduce .4375 to an equivalent vulgar fraction.

Ans. $\frac{7}{16}$.

III. *To find the value of a decimal in the known parts of the integer.*

RULE. Multiply the decimal by the number of the next lower denomination which makes one of the integer; then multiply the *decimal part* of the product, by the next lower denomination, and so on. The figures on the left of the separatrix, will express the value.

1. Find the value of .775 of a pound sterling.

$$\begin{array}{r} .775 \\ 20 \\ \hline 15.500 \\ 12 \\ \hline 6.000 \\ \hline \end{array}$$

Ans. 15s. 6d.

2. Find the value of .0125 lb. Troy.   Ans. 3 dwt.
3. Find the value of .625 cwt.   Ans. 2 qr. 14 lb.
4. Find the value of .009943 of a mile.

Ans. 17 yd. 1 ft. 5.9 in.

5. Find the value of .6875 yd.   Ans. 2 qr. 3 na.
6. Find the value of .3375 A.   Ans. 1 R. 14 r.
7. Find the value of .2083 hhd. of wine.

Ans. 13.1 gal.

8. Find the value of .785 bu.   Ans. 3 pk. 1 qt.+
9. Find the value of .625s.   Ans. 7½d.

IV. *To reduce the lower denominations of a compound number, to the decimal of a higher.*

RULE. Reduce the given quantity to a vulgar fraction by case VII. p. 56 ; and then reduce the vulgar fraction to its equivalent decimal.

Or, divide each denomination, beginning with the lowest, by its value in the next, and set the quotient on the right of the next higher denomination, with a decimal point between, then divide the next higher *together with its decimal*, in the same manner, and so on, as far as necessary.

1. Reduce 1 R. 14 r. to the decimal of an acre.

$$1 \text{ R. } 14 \text{ r.} = 54 \text{ r.} \qquad 40 \ | \ 14.$$
$$1 \text{ A.} = 160 \text{ r.} \qquad 4 \ | \ 1.35$$

$$\tfrac{54}{160} = .3375 \text{ Ans.} \qquad\qquad .3375.$$

2. What part of 1 cwt. is 2 qr. 14 lb. ?
     Ans. .625 cwt.

3. Reduce 17 yd. 1 ft. 5.988 in. to the decimal of a mile.
     Ans. .009943 m

4. Reduce 2 qr. 3 na. to the decimal of a yard.
     Ans. .6875 yd.

5. What part of a bushel, is 3 pk. 1.12 qt. ?
     Ans. .785 bu.

6. What part of 1 hhd. is 1.2 pt. ?
     Ans. .00238 hhd.

7. Reduce .26 d. to the decimal of a shilling.
     Ans. .02166+.

8. What part of £1 is 15s. 6d. ?   Ans. £.775.

9. What part of a pound Troy, is 5 oz. 11 dwt. 16 gr. ?
     Ans. .46944 lb.

10. Reduce .21 pt. to the decimal of a peck.
     Ans. .013125 pk.

11. Reduce 12s. 9d. to the decimal of a pound.
     Ans. .6375.

12. Reduce 3 pk. 3 qt. to the decimal of a bushel.
     Ans. .8593.

REDUCTION OF CURRENCIES.

Originally, the pound sterling was of the same value in the American colonies as in Great Britain, and a Spanish dollar worth 4s. 6d. But owing to the depreciation of the bills of credit emitted by the several colonies, the value of the pound, and consequently of its divisions, became various. Thus in New England, Virginia, Kentucky, Tenessee, 6s. are reckoned a dollar.

In New York and North Carolina, 8s. are a dollar.

In New Jersey, Delaware, Maryland, and Pennsylvania, 7s. 6d.

In South Carolina and Georgia, 4s. 8d.

In Canada and Nova Scotia, 5s.

At 6s. to the dollar, $\$1 = \frac{6}{20}$ or .3 of £1.

" 8s. - - - - $\$1 = \frac{8}{20}$ or .4 " £1.

" 7s. 6d.* - - $\$1 = \frac{3}{8}$ - - " £1.

" 4s. 8d. - - $\$1 = \frac{7}{30}$ - - " £1.

To reduce the several currencies to Federal Money.

RULE. Reduce the shillings, pence, &c., to the decimal of a pound; annex that decimal to the pounds, and divide by that fraction of £1, which makes a dollar in the given currency.

1. Reduce £45 15s. 7½d. New England currency to Federal Money.

$$
\begin{array}{r|l}
12 & 7.5 \\
20 & 15.6250 \\
.3 & 45.7812 \\
\end{array}
$$

Ans. $ 152.620.

2. Reduce £73 Virginia currency to Federal Money.

.3 | 73.0

Ans. $ 243.33⅓.

* 7s. 6d. = 90d.
£1   = 240d. $= \frac{3}{8}$ of £1.

3. £105 14s. 3¾d. New York currency.

12	3.75
20	14.3125
.4	105.7156

Ans. $ 264.289.

4. Reduce £125 8s. 6d. New Jersey currency to Federal Money.

12	6.
20	8.5

£125.425 ÷ ⅜ = $ 334.466+ Ans.

5. How many dollars in £25 3s. 7d Pennsylvania currency? Ans. $ 67.144.

6. How many dollars in 17s. 9d. ? Ans. $ 2.366.

7. Reduce £17 14s. 6d. South Carolina currency to Federal Money.

14s. 6d. = .725 of £1.

£17.725 ÷ $\frac{7}{30}$ = $ 75.965 Ans.

8. How many dollars in 11s. 6d. Georgia currency? Ans. $ 2 .464+.

9. Reduce £8 17s. 8d. Canada currency to dollars. Multiply the pounds by 4. Ans. $ 35.53.

10. How many dollars are 19s. Nova Scotia currency? Ans. $ 3.80.

NOTE. Whenever the sum to be reduced consists only of shillings, pence, &c., it is better to reduce the pence and farthings to the decimal of a shilling, and divide by the number of shillings in a dollar, if that be *even*.

11. Find the value in Federal Money of 17s. 6d. at 8s., 6s., and 5s., to the dollar.

$$17.5 \div \begin{cases} 8 = \$ 2.18\frac{3}{4} \\ 6 = \$ 2.91\frac{2}{3} \\ 5 = \$ 3.50 \end{cases}$$

12. How many dollars in 12s. 8¼d. in the currencies severally of New England, New York, and Nova Scotia ?

$$\text{Ans.} \begin{cases} \$ 2.118. \\ \$ 1.588. \\ \$ 2.541. \end{cases}$$

13. How many dollars in £5 15s. 4½d. of the currency of Great Britain ? Ans. $ 25.638.

The par value of a dollar is 4s. 6d. English currency.

$$\frac{4s.\ 6d. = 54d.}{£1 \quad = 240d.} = \frac{9}{40}.$$ Therefore, multiply by 40, and divide by 9.

14. What is the value in Federal Money of £1 English currency ? Ans. $ 4.44⁴⁄₉.

The method of reducing Federal Money to the currencies of the different states, Canada, and Great Britain, is to reverse the process directed in the rule, multiplying where we divided and dividing where we multiplied.

For examples, the learner may take the answers to the questions, under Reduction of Currencies, and reduce them back again.

MISCELLANEOUS EXAMPLES.

In the questions which follow, vulgar fractions, if they occur, are to be reduced to decimals ; as likewise, the inferior denominations of compound numbers.

1. How many yards of cloth in 4 remnants containing severally, 3 yd. 3 qr. ; 2 yd. 3 na. ; 1⅛ yd., and 2⅞ yd. ? Ans. 10.4375.

2. From 1 cwt., subtract 3 qr. 14¾ lb. Ans. .11839 cwt.

3. What is the cost of 3⅜ yd. of cloth at $ 5¾ per yd. Ans. $ 19.406+.

4. At $ 5¾ per yd., how much cloth can be purchased with $ 19.40625 ? Ans. 3.375 yd.

5. At $ 11.76 per cwt., what will ½ qr. of sugar come to ? Ans. $ 1.47.

6. How long must a laborer work, at the rate of $ .62½ per day, to earn $ 25 ? Ans. 40 days.

7. Travelling at the rate of 4⅗ miles an hour, in how many hours will a foot-man go 34½ miles ? Ans. 7.5 hours.

8. If 85 yards of cloth be bought for $ 191.25, and sold at $ 2.87½ per yard ; how much is the whole profit ? Ans. $ 53.12½.

9. How much butter at 9 cents a pound, will pay for 12 yards of cloth at $ 2.19 per yard? Ans. 292 lb.

10. At $ 45½ per acre, what is the value of 32 square rods of land? Ans. $ 9.10.

11. The loss of merchandise by a fire was estimated at $ 11372.75, ⅗ of which was insured; how much was the loss after deducting the insurance? Ans. $ 4549.10.

The loss here to the owner is ⅖ of the whole amount, and is best found by multiplying that by .4=⅖. With the vulgar fraction, both division and multiplication would have been necessary. In some cases, however, operating with the vulgar fraction is shortest.

Let it be observed, that the product of any number by a decimal, is always some *fractional part* of the multiplicand.

12. A benevolent individual whose income was $ 5000, devoted .12 of it to charitable objects; how much did he give away annually? Ans. $ 600.

13. $\frac{3}{25}$ of the capital stock of a bank, which was ½ a million of dollars, was owned by three individuals; what amount of stock had they? Ans. $ 60000.

14. If you add .05 of $ 759.06 to itself, what will be the amount? Ans. $ 797.013.

15. If I invest $ 5500 in the stock of an insurance company, and lose .08 of it; how much is my loss?

Ans. $ 440.

As a decimal of two places has 100 for its denominator, we may consider such fractions as expressing *per cent.*, in questions like the last: that is, I lost 8 per cent. (or $\frac{8}{100}$) of my investment.

16. How much is ¾ of $\frac{3}{25}$ of 786½.

Ans. 70.785.

17. At 3 cents a piece, how many oranges can be purchased with $ 11.46? Ans. 382.

18. At .07 per cent., how much capital must be invested to yield $ 602? Ans. $ 8600.

19. What is .07 of $ 8600? Ans. $ 602.

20. The children in a Sunday school contributed $ 5 to a charitable object, and it amounted to 6¼ cents a piece; how many children were there? Ans. 80.

21. Multiply 80 by .0625. Ans. 5.

It is evident from the last examples, that the quotient of a number divided by a decimal, is greater than the dividend, and the product of a number by a decimal is less than the multiplicand.

22. What is the sum of 19 and 5 hundredths ; 120 and 9 thousandths ; 36 ten thousandths ; and 464 and 8 millionths ?

23. From $\frac{85}{1000}$ take $\frac{96}{10000}$, and divide the difference by .002.   Ans. 37.7

24. The sum of $ 123.369 was to be paid by a certain number of persons, and it amounted to 3 dollars 9 mills a piece ; how many were there of them ?   Ans. 41 men.

25. Which is greatest, the product of 30 by .06, or its quotient by the same number ?   Ans. $\left\{\begin{array}{l}.1.8 \text{ product.} \\ 500. \text{ quotient}\end{array}\right.$

26. A lottery ticket of which A owned .4, drew a prize of $ 2500, what was his share of the prize ?   Ans. $ 1000.

27. $\left.\begin{array}{l}108. \\ 10.8 \\ 1.08 \\ .108\end{array}\right\} \div \left\{\begin{array}{l}12. \\ 1.2 \\ .12 \\ .12\end{array}\right.$ Each dividend is to be divided by *every* divisor.

28. Suppose the difference of two numbers to be 23.436, and the greater number 24 ; what is the less ?   Ans. .564.

29. The product of two numbers is 5, and one of the numbers 625 ; what is the other number ?   Ans. .008.

See the method of proof for multiplication.

30. How many rods, yards, &c. in .712 of a furlong ?
                              Ans. 28 r. 2 yd. 1 ft. 11 in.

QUESTIONS.

What does the term *decimal* mean ?
What is the use of the decimal point ?
In what respect do decimals differ from other fractions ?
Does a cipher on the right of a decimal increase or diminish its value ?
Why. and how much does a cipher *prefixed* to a decimal diminish its value ?
Why in addition and subtraction of decimals, must the decimal points stand under one another ?
Why is the process of adding and subtracting decimals, more simple and short, than the same operation with vulgar fractions ?
What is the effect on the value of a number, of removing the decimal

point toward the right ?   And what the effect of removing it toward the leaf ?

Why is the product of a number by a decimal less than the multiplicand ?

Why is the quotient greater than the dividend ?

DUODECIMALS.*

Duodecimals are fractional parts of a foot resulting from the division of it into 12 equal parts, and of each of those parts into twelfths, and so on.

The first divisions are called inches or primes ; the divisions of these, seconds ; then thirds, fourths, and fifths.

Hence the foot being taken as the integer.

$$1 \text{ prime} \quad (')= \tfrac{1}{12} \text{ of } 1 \text{ foot,}$$
$$1 \text{ second} \ ('')= \tfrac{1}{12} \text{ of } 1 \text{ prime,}$$
$$1 \text{ third} \quad (''')= \tfrac{1}{12} \text{ of } 1 \text{ second, \&c}$$

Duodecimals are applied to the measurement of surfaces and solids.

Th are added and subtracted like other compound numbers, but there is some peculiarity in the method of multiplying them into each other.

What are the superficial contents of a board 12 ft. 8 ′ long, and 2 ft. 2 ′ 2 ″ wide ?

As 2 feet is an integer, its product into any of the denominations of the multiplicand will evidently be the denominations themselves : but inches being only twelfths of a foot, their product into any denomination of the multiplicand can be but $\frac{1}{12}$ as much as if they were feet. And the product of seconds will be $\frac{1}{12}$ the product by the same number of primes.

---

* From the Latin numeral *duodecim,* signifying *twelve.*

	ft.	´	´´	´´´

Hence in multiplying by inches, we set the product one denomination lower than the product by feet, and the product of the seconds one denomination lower than that of primes, which diminishes each, in the ratio of the value of its multiplier to the next higher denomination.

	ft.			
	12	8		
	2	2	2	
	25	4		
	2	1	4	
		2	1	4
Ans.	27 ft.	7´	5´´	4´´´

RULE. Place the several terms of the multiplier under the corresponding terms of the multiplicand. Multiply first by the highest denomination, and set the product of each inferior denomination of the multiplier one place lower than the product of the next higher, always dividing by 12 when the product equals or exceeds that number.

What quantity of wood in a pile 9 ft. 7´ long, 3 ft. 8´ wide, and 4 ft. 3 ´ high ?

	ft.	´	´´	´´´
Length	9	7		
Width	3	8		
	28	9		
	6	4	8	
	35	1	8	
Height	4	3		
	140	6	8	
	8	9	5	0
Ans.	149 ft.	4´	1´´	0´´´

What quantity of surface in a floor 16 ft. 6 ´ long, and 12 ft. 9 ´ broad ?   Ans. 210 ft. 4 ´ 6 ´´.

How many feet in a board 14 ft. 10 ´ long, and 11 inches wide ?   Ans. 13 ft. 7 ´ 2 ´´.

When duodecimals are to be divided by one another, the inferior denominations must be reduced to the decimal of a foot.

If the square contents of a floor are 210 ft. 4′ 6″, and one of its sides 12 ft. 9′; what is the other side?

4′ 6″ =.375
9′   =.75   210.375÷12.75=16.5 or 16 ft. 6 in. Ans.

If a board 14 ft. 6′ long contain 13 ft. 3′ 6″, what is its width?   Ans. 11 inches.

As duodecimals can with equal convenience be reduced to decimals, it is unnecessary to multiply examples under this rule.

PRACTICE.

UNDER this head are included certain short practical methods of operating, applied for the most part to questions, where the value, or the quantity, in a lower denomination, is an even part of a higher denomination.

TABLE OF ALIQUOT (OR EVEN) PARTS.

Cents.	Parts of 1.	Months.	Parts of 1 year.	Parts of £1.	Parts of 1 shilling.	Parts of 1 cwt
50	$\frac{1}{2}$	6	$\frac{1}{2}$	10s. $=\frac{1}{2}$	6d. $=\frac{1}{2}$	56 lb.$=\frac{1}{2}$
33$\frac{1}{3}$	$\frac{1}{3}$	4	$\frac{1}{3}$	6s. 8d.$=\frac{1}{3}$	4d. $=\frac{1}{3}$	28 lb.$=\frac{1}{4}$
25	$\frac{1}{4}$	3	$\frac{1}{4}$	5s. $=\frac{1}{4}$	3d. $=\frac{1}{4}$	16 lb.$=\frac{1}{7}$
20	$\frac{1}{5}$	2	$\frac{1}{6}$	4s. $=\frac{1}{5}$	2d. $=\frac{1}{6}$	14 lb.$=\frac{1}{8}$
12$\frac{1}{2}$	$\frac{1}{8}$	1	$\frac{1}{12}$	3s. 4d.$=\frac{1}{6}$	1$\frac{1}{2}$d.$=\frac{1}{8}$	7 lb.$=\frac{1}{16}$
6$\frac{1}{4}$	$\frac{1}{16}$	or $\frac{1}{4}$ of	2s. 6d.$=\frac{1}{8}$	1d. $=\frac{1}{12}$	
5	$\frac{1}{20}$	3 mo.	1s. 8d.$=\frac{1}{12}$		

EXAMPLES.

1. At 1½d. per yard, what will 461 yd. of galloon come to?

The value at 1s. per yd.=461s. 6d.

$$
\begin{array}{cc}
 & \text{s.} \quad \text{d.} \\
1\tfrac{1}{2}\text{d.}=\tfrac{1}{8}\text{ of a shilling.} & \tfrac{1}{8})461 \quad 6
\end{array}
$$

Ans. 57 8¼=£2. 17s. 8¼d.

2. A. 3s. 8d. per bushel, what will 35½ bushels of wheat come to?

The value at £1 per bushel=£35 10s.

$$
\begin{array}{cc}
 & £ \quad \text{s.} \quad \text{d.} \\
6\text{s. }8\text{d.}=\tfrac{1}{3}\text{ of £1.} & \tfrac{1}{3})35 \quad 10
\end{array}
$$

Ans. £11 16s. 8d

3. At 15s. per yard, what is the cost of 56¾ yards?

$$
\begin{array}{cc}
15\text{s.}=\tfrac{3}{4}\text{ of £1.} & £ \quad \text{s.} \quad \text{d.} \\
\tfrac{3}{4}=\tfrac{1}{2}+\tfrac{1}{4}. & \tfrac{1}{2})56 \quad 15 \\
 & \tfrac{1}{2})28 \quad 7 \quad 6 \\
 & 14 \quad 3 \quad 9
\end{array}
$$

Ans. £42 11s. 3d.

4. At 75 cents a bushel, how much will 62½ bushels of barley come to?

$$
\begin{array}{cc}
 & \$ \quad \text{c.} \\
\text{At \$1 per bushel}= & 62.50 \\
.75=\tfrac{3}{4}=1-\tfrac{1}{4} & 15.62\tfrac{1}{2}
\end{array}
$$

Ans. $ 46.87½

5. At 25 cents a yard, what will 37¾ yards of cotton come to?

At $ 1 per yard= $ 37.75

25 cents=¼ $ 9.43¾ Ans.

6. At 16⅔ cts. per pound, what is the cost of 25¼ lb. of coffee?

$ c.

At $ 1 per pound= 25.25

16⅔=⅙. $ 4.20⅚ Ans.

7. At $ 9.58 per cwt. what will 1 cwt. 2 qr. 14 lb. of potash cost?

$ 9.58

2 qr.=½ 1 cwt. 4.79

14 lb.=¼ 2 qr. 1.19¾

Ans. $ 15.56¾

8. At $ 2.50 per cord, what will 1 cord 32 feet of wood cost?

$ 2.50

32 ft.=¼. .62½

Ans. $ 3.12½

9. At 87½ cents a pound, what is the cost of 1 lb. 2 oz. of tea?

$ .875

2 oz.=⅛ .109

Ans. $ .984

10. If the interest of a certain sum for a year be $ 18.78, what will be the interest for 10 months?

$$\$ 18.78$$

6 months $=\frac{1}{2}$ 1 year	9.39
4 months $=\frac{1}{3}$ 1 year	6.26

Ans. $ 15.65

---

11. What will 76 lb. brown sugar cost, at $7\frac{1}{2}$d. per lb. ?

	s. d.
At 1s. per pound $=$	76
6d. $=\frac{1}{2}$ of 1s.	38
$1\frac{1}{2}$d. $=\frac{1}{4}$ of 6d.	9 6

Ans. 47s 6 $=£2$ 7s. 6d.

The foregoing are a few examples of the various practical methods, which business men and accountants adopt for convenience and despatch of calculation.

Examples for Exercise.

12. 1 cwt. 3 qr. 14 lb. of raisins at £2 11s. 8d. per cwt :
Ans. £4 16s. $10\frac{1}{2}$d.
13. 1 cwt. 1 qr. 8 lb. of sugar at $ 8.65 per cwt. .
Ans. $ 11.42.
14. $362\frac{1}{4}$ bushels of wheat at $ $1.12\frac{1}{2}$ cents per bushel .
Ans. 407.53.
15. $27\frac{1}{2}$ gallons of brandy at $ 1.25 per gallon .
Ans. $ $34.37\frac{1}{2}$.
16. 60 bushels of apples at $16\frac{2}{3}$ cents per bushel ?
Ans. $ 10.
17. $75\frac{1}{2}$ bushels of potatoes at $33\frac{1}{3}$ cts. per bushel .
Ans. $ $25.16\frac{2}{3}$.
18. $46\frac{3}{8}$ lb. butter at $12\frac{1}{2}$ cts. per pound ?
Ans. $ $5.79\frac{3}{4}$.
19. 1 gal. 2 qt. 1 pt. of wine at $ $3.62\frac{1}{2}$ cts. per gal. .
Ans. $ 5.89.
20. 1 bu. 3 pk. 6 qt. of beans at $ $1.12\frac{1}{2}$ per bu. .
Ans. $ 2.18.

21. 29¾ yd. calico at 20 cts. per yd.　　　Ans. $ 5.95.
22. 27½ yd. silk at 9s. N. Y. currency per yd.

　　　　　　　　　　　　　　Ans. $ 30.93¾.
23. 1 cwt. 16 lb. iron at $ 6.75 per cwt.

　　　　　　　　　　　　　　Ans. $ 7.71¾.
24. 24 lb. sugar at $ 11.25 per cwt.　　　Ans. $ 2.41.
25. 1000 quills at ½ ct. a piece　　　　　Ans. $ 5.
26. The interest on a certain sum for a year being
$ 17.60, what is it for 7 m. 20 d.?

　　　　　　　　　　　　　　Ans. $ 11.24¼.
27. At 5s. 6d. per yd. N. Y. currency, what will 8 yd.
of gingham come to in Federal money?

As there are 8 shillings to the dollar, the cost of 8 yards
will be as many dollars as there are shillings in the price of
1 yd.; and the parts of a shilling, will be like parts of a dollar. 6d.=½ of 1 shilling. The answer therefore is $ 5.50.

28. At 4s. 3d. N. Y. currency, per gallon, what will 8
gallons of molasses cost?

　　　　　3d.=¼ of 1s.　　　　Ans. $ 4.25.

In New England currency, the price in shillings of 6
yards, pounds, &c., will be as many dollars as there are
shillings in the price of 1 yd., and the pence will be the
same aliquot parts of a dollar, as they are of a shilling.

29. At 1s. 6d. per bu. N. E. currency, what will 6 bu.
of oats cost in Federal money?

　　　　　　　　　　　　　　Ans. $ 1.50.
30. At 7s. 3d. per bu. N. Y. currency, what will 32 bu.
of wheat cost?

　　32=8×4　　8 bu.=$ 7.25×4=$ 29. Ans.

31. At 6s. 3d. per yd. N. Y. currency, what will 9 yds.
of Irish linen cost?

　　　　　　8 yd. =$ 6.25
　　　　　　1 yd.⅛=　.78⅛
　　　　　　　　　────────
　　　　　　Ans. $ 7.03⅛

(8*)

32. At 5s. 9d. per bu. (N. Y.) what will 25 bu. of corn come to ?

$$25 = 8 \times 3 + 1. \qquad \$ 5.75 \times 3 = 17.25$$
$$\tfrac{1}{8} = \qquad .71\tfrac{7}{8}$$

Ans. $ 17.96$\tfrac{7}{8}$

33. 15 yards of Irish linen at 4s. 4d. (N. E.) per yd. ?
                                   Ans. $ 10.83$\tfrac{1}{3}$.

34. Mr. Charles Lardner,

                    Bought of James Fisher.

1835.   s. d
Jan. 3.   9 yd. Bleached Cotton at 1   3   - - -
          6 lb. Souchong Tea   " 4   4   - - -
         12 "   Black   do.   " 3   9   - - -
         18 "   Coffee   " 1   3   - - -
          7 yd. Irish Linen   " 4   8   - - -
         24 "   Calico   " 1   8   - - -

                                        $ 29.57.

North Hampton, Jan. 3, 1835,
              Received payment,
                         JAMES FISHER.

35. Mr. John Schoolcraft,

                    To C. P. Smith,   Dr.

1836.   s. d.
Jan. 8. To 10 lb. Coffee   at 1   9   - - -
  "   " 12 " Sugar   " 1   3   - - -
  " 12.   " 18 " Salmon   "   6   - - -
Feb. 1.   "   9 " Black Tea " 5   3   - - -
  "   " 24 " Buckwheat "   3   - - -
  " 6.   "   9 " Corn'd Beef "   4   - - -

                                        $ 12.21.

New York, March 4, 1836.
              Received payment,
                         C. P. SMITH.

COMPARISON OF NUMBERS.

THE part or parts, which one number is of another, is expressed by making the former the numerator, and the latter the denominator of a fraction. Thus, if we compare 2 and 3; 2 is $\frac{2}{3}$ of 3, 3 is $\frac{3}{2}$ of 2.

Again, if we compare 4 and 6; 4 is $\frac{4}{6}$ of 6, and 6 is $\frac{6}{4}$ of 4. These fractions reduced, become $\frac{2}{3}$ and $\frac{3}{2}$; the relation of the terms remaining unchanged.

In order to compare numbers in this manner, they must be *in the same denomination.*

If, for example, we would express the part which 5 feet is of 2 yards, it will not do to say, $\frac{5}{2}$ of 2 yards; but reducing the 2 yards to feet, we shall have a true comparison, namely $\frac{5}{6}$ of 2 yards.

If the numbers compared contain decimals, the decimal places must be made equal, and if they contain vulgar fractions, these must be reduced to a common denominator, and their numerators be compared.

1. What part of \$1 is 60 cents?

$$\frac{.60}{1.00} \text{ or } \frac{3}{5} \text{ Ans.}$$

2. What part of $\frac{2}{3}$ is $\frac{1}{4}$?

$$\frac{2}{3}=\frac{8}{12}$$
$$\frac{1}{4}=\frac{3}{12} \qquad\qquad \text{Ans. } \frac{3}{8}.$$

3. What part of $\frac{1}{4}$ is $\frac{2}{3}$?   Ans. $\frac{8}{3}$.
4. What part of 7 is 2?   What part of 2 is 7?
5. What part of 12 is 9?   What part of 9 is 12?
6. What part of 3 cwt. is 1 cwt. 3 qr.?   Ans. $\frac{7}{12}$.
7. What part of 6s. is 8s. 6d.?   Ans. $\frac{51}{36}$.
8. What part of 5.6 is .56?   Ans. $\frac{1}{10}$.
9. What part of 3 gal. is 2 qt. 1 pt.?   Ans. $\frac{5}{24}$.

The relations expressed by numbers compared in this way, may be applied to the solution of a variety of practical questions.

10. If 3 men can build 7 rods of wall in a day, how many rods can 9 men build?

9 is $\frac{9}{3}$ of 3. $\frac{9}{3}$ of $7=7\times3=21$ rods, Ans.

11. If 3 horses eat 8 bushels of oats in 2 weeks, how long will it take them to eat 40 bushels?

There will be the same relation between the periods of time, that there is between the quantities consumed.

40 is $\frac{40}{8}$ of 8. $\frac{40}{8}$ of 2 = 2×5 = 10 weeks, Ans.

12. If $4\frac{1}{2}$ tons of hay will feed 3 cattle over the winter, how many tons will feed 25 cattle?

25 cattle will consume $\frac{25}{3}$ as much as 3 cattle.
4.5 ÷ 3 = 1.5. 1.5 × 25 = $37\frac{1}{2}$ tons, Ans.

13. If 19 gallons of molasses cost $ 11 what will $3\frac{4}{5}$ quarts cost?

In order to compare these two quantities, they must be brought to the *same denomination*. 19 gal. = 76 qt. $3\frac{4}{5}$ = 3.8 qt. Adding a cipher to 76 to balance the decimal place in 3.8, the relation of the quantities will be

$$\frac{38}{760} = \frac{1}{20}. \qquad 11 ÷ 20 = .55 \text{ cts. Ans.}$$

From the foregoing examples it will be seen, that if the answer sought is more than the given term, (of the same kind,) the greater of the other terms is made the numerator; and the less of the other terms the denominator of the fraction, by which the comparison is instituted.

When either of the terms contains a vulgar fraction, it may be reduced to its equivalent decimal; the *decimal places in all cases being made equal* in the two terms.

14. If 16 bushels of oats cost $ 6.75, what will 320 bu. cost? Ans. $ 135.

15. If 3 yards of cloth may be bought for $ 12.75, how many yards may be bought for $ 102? Ans. 24.

16. If 1 acre and 20 rods of ground produce 45 bu. of wheat; at that rate, how much will 9 acres produce? Ans. 360 bu.

17. If $1\frac{3}{4}$ yd. of cloth be worth $ 5, what is $\frac{5}{8}$ of a yard worth? Ans. $ 1.78$\frac{4}{7}$.

18. Boarding at 12s. 6d. per week; how long will £32 10s. last me? Ans. 52 weeks.

19. If a barrel of beef last 10 men 95 days, how long will it last 25 men? Ans. 38 days.

20. If I lend a man $ 200 for 60 days, how long ought he to lend me $ 275 to requite the favor?

Ans. $43\frac{7}{11}$ days.

21. How many years will it require for 5 cents to gain the same interest, that $ 100 does in 1 year?

Ans. 2000 years.

22. If it be required to line cloth $\frac{7}{8}$ of a yd. wide, with lining $\frac{3}{4}$ wide; what must be the relative quantity of the lining?

Ans. $\frac{7}{6}$.

PROPORTION.

FROM the comparison of numbers we derive their ratio.

Ratio is the mutual relation of two numbers in respect to quantity. It is expressed by making one of the numbers the numerator, and the other the denominator of a fraction. Thus the ratio of 3 to 6 is $\frac{6}{3}$ or 2, and the inverse ratio $\frac{3}{6}$ (or $\frac{1}{2}$.) The two expressions are equivalent, and import that one number is double the other.

Two numbers thus compared are called a couplet; and the first is named the antecedent; the second, the consequent of the couplet. Both are called terms.

If two or more couplets of numbers have equal ratios—that is, if the quotient of the second divided by the first, is equal to the quotient of the fourth by the third, or inversely—the numbers are said to be *proportional*. Thus, 2, 4 3, 6, are proportional; that is $\frac{4}{2}=\frac{6}{3}$ and $\frac{2}{4}=\frac{3}{6}$.

A proportion is usually indicated by points, thus:— 2 : 4 : : 3 : 6, which is read 2 is to 4 as 3 is to 6.*

---

* When 4 numbers constitute a proportion, they are also a proportion when taken inversely, or alternately. Thus taking the same example :

Directly   3 : 6 : : 4 : 8
Inversely   6 : 3 : : 8 : 4
Alternately 3 : 4 : : 6 : 8

Other changes may be made in the order of the terms, without affecting the equality of the ratios.

When four numbers are proportional, the product of the two means, or middle terms, is equal to the product of the two extremes or outer terms. Thus in the proportion above, $2\times6=3\times4$.

The reason of this equality may be explained as follows : In every proportion, the consequent of each of the two couplets is the product of its antecedent by the same number, that is, the ratio. Thus in the proportion $3 : 6 : : 4 : 8$; 6 is the product of 3 multiplied by 2, and 8 the product of 4 multiplied by 2. And if the proportion be taken inversely, $6 : 3 : : 8 : 4$, the same is true of it. The ratio is then $\frac{3}{6}$ or $\frac{1}{2}$, and $6\times\frac{1}{2}=3$, and $8\times\frac{1}{2}=4$.

Now each of the consequents can in all cases be resolved into 2 factors, one of which is the ratio, and the other its own antecedent. Thus in the proportion

$$3 : 6 : : 4 : 8. \qquad 6=(3\times2) \text{ and } 8=(4\times2)$$

and the proportion may be stated :

$$3 : (3\times2) : : 4 : (4\times2); \text{ or inversely,}$$
$$6 : (6\times\tfrac{1}{2}) : : 8 : (8\times\tfrac{1}{2},)$$

The first term is therefore, a factor in the second, and the third term a factor in the fourth, and the ratio, from the very definition of proportion, is common to the two consequents. To say then, that the product of the two means is equal to the product of the two extremes, is merely asserting that the products of the same factors are equal.

This equality of the product of the means to that of the extremes, enables us, by having any three terms of a proportion, to find the fourth. For the product of any two numbers divided by one of the numbers, gives us the other as a quotient. Therefore, if we have the first three terms of a proportion, the product of the second and third divided by the first will give the fourth. Taking for example the proportion $3 : 6 : : 4$ $6\times4=24\div3=8$ the fourth term.

Now the terms of a proportion, which we have treated as abstract numbers, may be employed to represent things ; and the principles stated above, be applied to the solution of a vast variety of practical questions. Suppose, for illustration, the cost of 3 yards of cloth to be 12 dollars, and we wish to find the cost of 14 yards at the same rate. The ratio between 3 yards and 14 yards will evidently be the same, as the ratio between 12 dollars and the cost of 14 yards, which we wish to find. We may, therefore, make a proportion, and say

yd. yd. $

3 : 14 : : 12 : A.

Here we have the two middle terms and one of the extremes given, to find the other extreme. We have learned, that the product of the two means is equal to the product of the two extremes. We may, therefore, substitute this product; and, dividing by the given extreme, find the other. Thus, $14 \times 12 = 168 \div 3 = \$ 56$ Answer.

Again, If a perpendicular staff 4 feet high, cast a shadow 5 feet long, what is the height of a tower, which at the same time casts a shadow 140 feet in length?

There must be the same ratio between the heights of the objects, that there is between the lengths of their shadows.

s. s. h

Therefore 5 : 140 : : 4 : A.

$140 \times 4 = 560 \div 5 = 112$ ft. Ans.

In every proportion, if the second term be greater than the first, it is evident that the fourth will be greater than the third; if the second be less than the first, the fourth will be less than the third.

In order that the ratio between two terms be a true one, the numbers denoting them *must be of the same denomination.*

If 4 yards of cloth cost 32 dollars, what will 3 quarters of a yard cost?

The true ratio between the two pieces of cloth, is not that of 4 to 3, but (reducing the yards) 16 to 3.

qr. qr. $

Therefore, 16 : 3 : : 32 : A.

$32 \times 3 = 96 \div 16 = \$ 6$ Ans.

The nature and properties of a proportion are the foundation of what—from its great utility—is termed the GOLDEN RULE; and from the circumstance that there are three given terms—

THE RULE OF THREE.

The object of this rule is, from having three terms given, to find a fourth, to which the third shall have the same ratio which the first has to the second.

Two of the given numbers are always of the same kind, one of which must be the first term in stating, and the other the second. Of these two numbers, one is a supposition, and has commonly the word *if*, or some like expression connected with it ;—the other is a demand, and is the thing to be *found out.* The third number is of the same kind with the answer sought; and must possess the third place in the statement. The following is a question under this rule.

If 5 yards of cloth cost $ 9, what will 20 yards cost ?

Here, the numbers that are of the same kind are 5 yards and 20 yards : they are both *cloth.* Moreover one of them is preceded by a *supposition:* " *If* 5 yards cost," and the other is the question or thing to be found out. " What will 20 yards cost ?"

Again, the remaining number (9) is of the same kind as the answer sought : it is *money.*

Now it is evident, that there should be the same ratio between 9 dollars and the answer to the question, that there is between 5 yards and 20 yards. Therefore it is stated

$$\text{yd. yd.} \qquad \$$$
$$5 : 20 : : 9 : A.$$

RULE. State the question by placing the number which is of the same kind with the answer sought, for the third term ; and the two numbers which are of the same kind with each other for the first and second terms, in such order that the first shall have the same ratio to the second term, as the third has to the answer.

2. Bring the first and second terms to the same denomination, and the third to the lowest denomination mentioned in it.*

---

* It is often better to reduce the lower denominations to the decimal of the highest.

3. Multiply the second and third terms together, and divide the product by the first, and the quotient will be the answer, in that denomination which the third term was left in.

In arranging the first two terms, we have only to consider, whether the answer ought to be greater or less than the third term. If greater, then the greater of the two terms should possess the second place; but if less, then the less of the two like terms should possess the second place.

Thus in the question : If 5 yards cost $ 9, what will 20 cost, it is evident that the answer ought to be more than 9 dollars ; and we arrange the terms :

yd. yd. $
 5 : 20 : : 9 : A. $20 \times 9 = 180 \div 5 = $ 36$ Ans.

Had the question been : If 20 yards cost $ 36, how many dollars will 5 yards cost ? the answer required would be less than the third term, and the statement would stand :

yd. yd. $
 20 : 5 : : 36 : A $36 \times 5 = 180 \div 20 = $ 9$ Ans.

1. If 365 men consume 75 cwt. of pork in 9 months, how much will 500 men consume in the same time ?

$$
\begin{array}{ccc}
\text{m.} & \text{m.} & \text{cwt.} \\
365 : & 500 : : & 75 \\
& & 75
\end{array}
$$

365)37500(102$\frac{54}{73}$ Ans.
   365
  ————
  1000
   730
  ————
  $\frac{270}{365} = \frac{54}{73}$.

After the division we have a remainder 270. Placing the divisor under it as a denominator, we have the fraction $\frac{270}{365}$* $= \frac{54}{73}$ and the answer is 102$\frac{54}{73}$ cwt.

---

* Fractions should in all cases be reduced to the *lowest terms.*

If we wish to obtain the value of this fraction in the known parts of the integer, (that is of 1 cwt.) we multiply the numerator by the tabular parts of the integer, and divide by the denominator, as directed in case VI. of Vulgar Fractions, page 55.

```
        54
         4
       ----
   73)216(2
      146
       ----
        70
        28
       ----
   73)1960(26
      146
      ----
       500
       438
      ----
        62
```

Ans. 102 cwt. 2 qr. 26$\frac{62}{73}$ lb.

We might have added ciphers to the remainder, and have obtained a decimal expression of the fractional parts.

```
   365)270.0000(.7397
       2555
       ----
       1450
       1095
       ----
       3550
       3285
       ----
       2650
       2555
       ----
```

The value of this decimal is found by case III. page 74.

Ans. 102.7397 cwt.

Operations under this rule may often be considerably abridged, by dividing the first term, and either of the other two terms, by any number which will divide them without remainder.

2. If 12 gallons of wine be worth $ 30, what is the value of a cask of the same wine containing 31$\frac{1}{2}$ gallons ?

$$\text{g.} \qquad \text{g.} \qquad \$$$
$$12 : 31.5 :: 30 : A.$$

Dividing the first and
third terms by 6.

$$2 : 31.5 :: 5$$
$$5$$
$$\overline{}$$
$$2)157.5$$
$$\overline{}$$

Ans. $ 78.75

3. If a family of 12 persons consume 3 barrels of flour in 2 months, how many barrels will serve them when there are 36 in the family?

$$\text{p.} \qquad \text{p.} \qquad \text{bl.}$$
$$12 : 36 :: 3 : A.$$

Dividing by 12 - - $1 : 3 :: 3$

$$3 \times 3 = 9 \text{ bl. Ans.}$$

4. If 48 yards of cloth cost $ 67.25, what will 160 yards cost at the same rate?

$$\text{yd.} \qquad \text{yd.}$$
$$48 : 160 :: 67.25 : A.$$

Dividing by 16 $\quad 3 : 10 :: 67.25$

Ans. $ 224.166.

5. What will 11 cwt. 1 qr. of sugar come to, if £8 8s. are paid for 3 cwt?

cwt. cwt. qr.   £ s.
$$3 : 11 \ 1 :: 8 \ 8 : A.$$
$$4 \quad 4 \qquad 20$$
$$\overline{} \quad \overline{} \qquad \overline{}$$
$$12 : 45 :: 168$$
$$45$$
$$\overline{}$$
$$840$$
$$672$$
$$\overline{}$$
$$12)7560$$
$$\overline{}$$

Dividing the first and
third terms by 12, we have
$$1 : 45 :: 14$$
$$14$$
$$\overline{}$$
$$180$$
$$45$$
$$\overline{}$$
630s.=£31 10.

630s.=£31 10s. Ans.

(9)

Instead of reducing the first and second terms to quarters, it would be better to reduce 1 qr. to the decimal of 1 cwt., and the 8s. to the decimal of a pound.

1 qr.＝.25
8s.＝.4　　　　　　　3 : 11.25 : : 8.4 : A.
Dividing by 3　　　　1 : 11.25 : : 2.8
　　　　　　　　　　　　　2.8
　　　　　　　　　　　　————
　　　　　　　　　　　·9000
　　　　　　　　　　　2250 .

Ans. £31.5

6. If 12 horses consume 42 bushels of oats in 3 weeks, what quantity will serve 20 horses the same time ?
Ans. 70 bushels.

7. If 70 bushels or oats will serve 20 horses for 3 weeks, how many horses will 42 bushels serve for the same time ?
Ans. 12 horses.

8. If 42 bushels of oats will last 12 horses 3 weeks, how long will 70 bushels of oats last them ? Ans. 5 weeks.

9. If 76 yards of cloth cost $ 136.80, how much is it per ell English ? Ans. $ 2.25.

10. At $2.25 per ell E., how many yards of cloth can be bought for $ 136.80 ? Ans. 76 yards.

11. If 12 bushels of wheat be bought for $ 13.32, how many can be bought with $ 51.06 ? Ans. 46 bushels.

12. If 6 bushels of corn cost $ 4.75, what will 75 bushels cost ? Ans. $ 59.37½.

13. An insolvent debtor fails for $ 13746.75, of which he is able to pay only $9164.50; how much does he pay on a dollar, and what will A receive to whom he owes $ 2139 ?
Ans. $\left\{\begin{array}{l}66\frac{2}{3}\text{ cts. on }\$ 1.\\ \$ 1426.\end{array}\right.$

14. If 18 barrels of flour cost $ 99, what will 73 barrels cost at the same rate ? Ans. $ 401.50.

15. In its diurnal revolution, the earth moves through 15 degrees of a circle in 1 hour; how many degrees will it move through in 24 minutes ? Ans. 6 degrees.

16. If 15 degrees of longitude cause a difference in time of 1 hour, what will be the hour at Washington, when it is 12 o'clock at a place situated 36° 45' east of it?

Ans. 9 o'clock 33 minutes.

17. If a man's yearly income be $2000, and his average expenditures be $25.37½ per week, how much can he lay up in 6 years, including one Leap-Year?

Ans. $4057.62¼.

18. If a perpendicular post 6 feet 3 inches high, cast a shadow 8 feet 9 inches in length, what is the height of a steeple which at the same time casts a shadow 133 feet?

Ans. 95 feet.

19. If 7 lb. of sugar cost 75 cents, how many pounds can be bought with $9? Ans. 84 lb.

20. At the rate of 9 yards for £5 12s., how many yards of cloth can be bought for £44 16s.? Ans. 72 yd.

21. At half a guinea per week, how long can I be boarded for 20 pounds sterling? Ans. 38$\frac{2}{21}$ weeks.

22. If 240 bushels of wheat are purchased at the rate of $22½ for 18 bushels, and sold at the rate of $33¾ for 22½ bushels, what is the profit on the whole? Ans. $60.

23. If $100 gain $7 interest in a year, what will $49.75 gain in the same time? Ans. $3.48¼

24. If $49.75 gain $3.48¼ in a year, what principal will gain $7 in the same time? Ans. $100.

25. If 7 yards of cloth cost $15.47, what will 12 yards cost? Ans. $26.52.

26. What will 26 yd. of cloth come to, if $6.90 are paid for 13 ells French?

NOTE. Instead of multiplying and dividing the terms in full, it is generally practicable to shorten the operation by the method explained on page 31.

If there are more decimals on one side of the line than on the other, they may be balanced by adding ciphers to *any* number on the opposite side; or an inverted comma may be used to distinguish them from integers.

Taking for illustration the last question; the statement stands:

e.   yd. $

13 : 26 : : 6.90 : A.

The first and second terms are to be reduced to quarters; we may, however, merely set down the multipliers, with a sign of multiplication, and reduce the terms by division, as before directed.

$$\text{Dividend} \quad \frac{26.4.6'90}{13.6} = \frac{2.4.1'15}{1.1} = \$\, 9.20 \text{ Ans.}$$
$$\text{Divisors}$$

This method is particularly convenient where the terms of the statement involve vulgar fractions. In that case, we have only to place the denominators on the opposite side of the horizontal line to their numerators, and the whole may be treated as integers.

27. If $\frac{3}{8}$ of a yard of velvet cost $\frac{2}{5}$ of a pound sterling, what will $\frac{5}{16}$ of a yard cost?

yd. yd. £
$\frac{3}{8} : \frac{5}{16} : : \frac{2}{5} :$ A. $\quad$ Dividend $\frac{8.5.2}{3.16.5} = \frac{1}{3} = 6s.\, 8.$ Ans.
$\qquad\qquad\qquad\quad$ Divisors

The first term being the divisor, its numerator is placed below the line : the second and third terms retain their position. In this example, the 5 cancels the 5; 16 being divided by 8 gives a quotient 2, which is balanced by 2 above the line. The terms thus reduced stand : $\frac{1.1.1}{3.1.1}$

Units below the line, may in all cases be dropped, as they do not affect the result.*

28. If $\frac{5}{8}$ of a yard cost $\frac{3}{7}$ of a pound, what will $\frac{2}{5}$ of an ell English cost?

yd. e. £
$\frac{5}{8} : \frac{2}{5} : : \frac{3}{7} :$ A. $\quad \frac{8.3.5.3}{5.4.5.7} = \frac{18}{35} = 10s.\, 3\frac{1}{4}d.$

In this example, $\frac{5}{8}$ of a yard and $\frac{2}{5}$ of an ell are to be reduced to the same denomination, by case VIII. page 57. This is done by multiplying their numerators by 4 and 5 respectively. This reduces them to the fraction of a quarter : that is, $\frac{5}{8}$ yd. $= \frac{20}{8}$ qr. and $\frac{2}{5}$ e. $= \frac{10}{5}$ qr. We therefore, place

---

* When any term is reduced, a pencil may be drawn across it, and its quotient be set above or below it.

the multipliers next their respective terms, with the sign of multiplication between, and then reduce all the numbers by division.

29. If $1\frac{3}{4}$ yd. of cloth cost 42 cents, what will $87\frac{1}{2}$ yards cost ?

yd.   yd.

$1\frac{3}{4}$: $87\frac{1}{2}$: : 42 : A.

$$\frac{4 \cdot 175 \cdot {}^{\prime}42}{7 \cdot 2} = 2 \times 25 \times .42 = \$21 \text{ Ans.}$$

30. If $8\frac{3}{4}$ gallons of molasses cost $ 4.20, what will $13\frac{4}{7}$ gallons cost ?

$8\frac{3}{4} = \frac{35}{4}$ Dividend $\dfrac{4 \cdot 9\,5 \cdot 4{}^{\prime}20}{35 \cdot 7}$

$13\frac{4}{7} = \frac{95}{7}$ Divisors $\quad = \dfrac{4 \cdot 19 \cdot {}^{\prime}60}{7} = \$ 6.51$ Ans.

NOTE. Whenever any of the terms of a statement contain a vulgar fraction, it may be reduced to its equivalent decimal. Which of the two methods to adopt, is a matter to be determined by convenience, and the judgment of the scholar.

In the foregoing example, if we convert the fractions to decimals, the terms will stand 8.75 : 13.57142 : : 4.20. The first method is preferable in this instance.

31. If a piece of land of a certain length, and 4 rods in breadth, contained $\frac{3}{4}$ of an acre ; how much would there be, if it were $11\frac{3}{5}$ rods wide ?   Ans. 2 A. 28 rods.

32. If $8\frac{2}{5}$ lb. of tobacco cost $ $1\frac{3}{4}$, what quantity can be bought for $ 317.23 ?   Ans. 13 cwt. 2 qr. $10\frac{3}{5}$ lb.

33. If $15\frac{5}{8}$ bushels of clover-seed cost $ $156\frac{1}{4}$, what quantity can be bought for $ $95\frac{3}{4}$?   Ans. 9 bu. 2 pk. $2\frac{2}{5}$ qt.

34. If $\frac{2}{5}$ of an ell English cost $\frac{1}{3}$ of a pound, what will $12\frac{1}{2}$ yards cost ?   Ans. £5 11s. $1\frac{1}{3}$d.

35. If $12\frac{1}{2}$ cwt. of iron cost $ $42\frac{1}{4}$, what will $48\frac{3}{8}$ cwt. cost ?   Ans. $ 163.50.

36. If $6\frac{1}{2}$ bushels of oats cost $ 3, what will $9\frac{1}{4}$ bu. cost ?   Ans. $ 4.269.

37. A farmer sold 17 bushels of barley and 13 bushels of wheat, for $ 31.55, the wheat at 35 cents a bushel more than the barley ; what was the price of each per bushel ?

    Ans. $\left\{\begin{array}{l}\text{Barley } \$ .90. \\ \text{Wheat } \$ 1.25.\end{array}\right.$

38. If 12 men can build a wall in 20 days, how many men can do the same in 8 days?

This question belongs to what is commonly called "Inverse Proportion." Those who adopt this distinction, would state it,

$$\begin{array}{ccc} d. & m. & d. \\ 20 : 12 : & : 8 : A. \end{array}$$

and then multiply the *first* and *second* terms together. But the terms thus arranged, evidently do not constitute a true proportion. The required term, must obviously be greater than 8; but 12 is less than 20. Moreover, the product of the means is not—as in all true proportions—equal to the product of the extremes. As the distinction of inverse proportion is both unscientific and unnecessary, it is now pretty generally discarded. The general rule is applicable to all questions which can arise under proportion, according to which, the foregoing example is stated thus:

$$\begin{array}{ccc} d. & d. & m. \\ 8 : 20 : & : 12 : A. \end{array}$$

Multiplying the second and third terms together, and dividing by the first we shall have the answer; 30 men.

There can be no difficulty in determining the proper order of the two like terms, if we consider, whether the answer ought to be greater or less than the third.

39. If a man perform a journey in 5 days, when the day is 12 hours long, in how many days will he perform it when the day is 10 hours long?

Here, the shorter the day, the more days it will require to perform the same journey; consequently the statement stands,   10 : 12 : : 5 : A.   Ans. 6 days.

40. What number of men will it require to execute in 2 months, a work which it would require 120 workmen to do in 8 months?   Ans. 480.

41. A family of 8 persons have provisions to subsist them a year; if at the end of 6 months 2 of the family leave, how long will their supply last them?   Ans. 14 months.

42. If A can mow an acre of grass in 6 hours, and B in 8 hours; how much will they jointly mow in 10 hours?

Ans. $2\frac{11}{12}$ acres.

43. If when flour is $ 6.50 a barrel, the bakers' loaf should weigh 10 oz., how much ought it to weigh when flour is $ 7.62$\frac{1}{2}$? Ans. 8$\frac{1}{2}$+ oz.

44. If 20 men can perform a piece of work in 15 days, how many men must be added to the number, that it may be accomplished in $\frac{4}{5}$ of the time? Ans. 5 men.

45. What quantity of water added to 31$\frac{1}{2}$ gallons of whiskey, which cost $ 13.50, would enable the purchaser to sell it for 40 cents per gallon, at a profit of 10 cents a gallon on his purchase? Ans. 10$\frac{1}{8}$ gallons.

46. If it require 90 yards of carpeting $\frac{3}{4}$ wide to carpet a floor, how many yards 1$\frac{1}{8}$ wide would be sufficient?

Ans. 60 yards.

47. If a courier, travelling 13$\frac{5}{8}$ hours a day, perform a journey in 35$\frac{1}{2}$ days, how long will it require, if he travel but 11$\frac{9}{10}$ hours a day? Ans. 40 days, 15$\frac{1}{2}$+h.

48. Suppose 650 men in a garrison, have provision sufficient to last them 2 months, how many men must leave, in order that the provision may last the residue 5 months?

Ans. 390 men.

49. At what time between the hours of 8 and 9, are the hour and minute hands of a clock together?

Ans. 8 o'clock, 43 min. 38$\frac{2}{11}$ sec.

50. If A can mow $\frac{1}{5}$, B $\frac{1}{6}$, and C $\frac{1}{8}$ of an acre of grass in an hour, in how many hours can they together mow 3$\frac{1}{3}$ acres? Ans. 6 h. 30 m. 30$\frac{1}{2}$ sec.

51. A man and two boys are employed in hoeing a field of corn; the man is able to hoe 3 rows while each of the boys hoes 2: if the former could perform the whole work alone in 8 days, how long will it take him with the assistance of the boys? Ans. 3$\frac{3}{7}$ days.

52. If by working 6$\frac{3}{5}$ hours a day, a man can accomplish a job in 12$\frac{1}{2}$ days, how many days will be required, if he work 8$\frac{1}{3}$ hours a day? Ans. 9$\frac{9}{10}$ days.

53. In exchange for 120 bushels of wheat valued at $ 1$\frac{1}{2}$ per bushel, A receives of B $ 65 in cash, and the balance in oats at 40 cents a bushel: what quantity of oats does he receive? Ans. 287$\frac{1}{2}$ bushels.

54. If A and B together can do a piece of work in 7 days, and B alone in 12 days, in how many days can A alone do $\frac{2}{3}$ of it? Ans. 11$\frac{1}{5}$ days.

COMPOUND PROPORTION.

By this rule, questions which in simple proportion would require two or more statements are reduced to a single statement.

EXAMPLE.

If £100 in 12 months gain £6, what will £400 gain in 7 months ? .

The result in this question, depends upon two considerations ; the amount of money at interest, and the length of time it is continued at interest.

In regard to the first, the statement would be,

prin. prin.   in.
100 : 400 : : 6—to £24, the interest of £400 for 12 months.

The consideration of time would require the statement to be,

m. m.   £
12 : 7 : : 24—to £14 the interest of £400 for 7 months-

Here it is to be observed, that the quotient arising from dividing the *product* of the two last terms of the first statement by the first term, is *in all cases* to be made the third term of the subsequent statement, and to be multiplied into its middle term.   If then we multiply that *product* by this middle term *before* division, the result will be the same. But this result is to be divided by the first term of the last statement.   It will have the same effect, however, to increase the first divisor (before the first division) as many times as there are units in the second divisor.

Hence, the proportions may be combined in the following manner :

$$100 : 400 \brace 12 : 7 \quad : : 6*$$

Then $\overline{400 \times 7 \times 6} \div \overline{100 \times 12} = £14$ Ans.

---

* It will be observed, that the ratio of the third term to the answer, is not the same as that of either of the two couplets.   There are in fact,— as explained above—two distinct proportions ; the combination of which is a matter of convenience, rather than of scientific accuracy.

RULE. Make that number which is of the same kind with the answer, the third term ; and of the remaining numbers, compare any two that are of a kind, and place them on the left of the third term, in such order, that their ratio may accord with the conditions of the question. Then divide the continued product of all the middle terms, and the third term, by the product of the first terms.

The method of statement will best be illustrated by an example.

1. If 8 men can build a wall 20 feet long, 6 feet high, and 4 feet thick in 12 days, in what time will 24 men build one 200 feet long, 8 feet high, and 6 feet thick ?

Here the answer sought is days. We, therefore, place 12 days for the third term, and comparing the number of men, it is evident that 24 men will re-

men	24 :	8	
length	20 :	200	d.
height	6 :	8	: : 12
thickness	4 :	6	

quire less time to do the same piece of work than 8 men. We, therefore, place the less of the two numbers last. We next compare the length of the two walls, and see that one 200 feet long, will require more days than one 20 feet long, and place the longest last. For the same reason, we arrange the heights of the two walls in the same order ; and so also of the thickness. The continued product of all the middle terms and the third term, divided by the product of all the first terms gives the answer 80 days.

But the labor of multiplication (as in simple proportion) may be shortened by reducing any of the first terms, and any of the other terms proportionally by division.

Therefore the numbers whose products are to form the dividend, may be placed above a line with a sign of multiplication between them, and those forming the divisor below, and the whole be reduced by the method already shown.

$$\frac{\text{Dividend } 8\times200\times8\times6\times12}{\text{Divisors } 24\times20\times6\times4} = \frac{10\times2\times12^*}{3} = 80 \text{ days Ans.}$$

By the common method.   200
                          8
                        ─────

24	1600
20	8
─────	───────
480	12800
6	6
─────	───────
2880	76800
4	12
─────	───────

11520) 921600(80 days.
        92160
      ─────────

2. If 20 bushels of wheat are sufficient for a family of 8 persons 5 months, how many bushels will be sufficient for 4 persons 12 months ?

Persons 8 : 4 ⎱
Months 5 : 12 ⎰ ∵ : 20 bu.

The question here is " how many *bushels*," &c., and the answer is *grain*. Consequently, the last term in the statement must be grain. By the *supposition*, 8 persons ate 20 bushels, and it is evident that 4 persons would require a less quantity for the same time : therefore, in regard to the *number* of persons, the answer should be less· than 20 bushels. But, the 8 persons were only 5 months in consuming the 20 bushels, whereas the 4 are to be fed 12

* The quotients, instead of being carried out in this manner, may just as well be set above or below the respective terms, and a pencil be drawn over the latter to show that they are disposed of.

If a point be used as the sign of multiplication, decimals may be denoted by the inverted commas ; or they may be balanced by an equal number of decimal places annexed to *any* term on the opposite side of the line.

months; therefore, in reference to *time*, the answer ought to be more than 20 bushels, and the greater of the two terms is placed last.

3. If 240 men in 5 days of 10 hours each, can dig a trench 230 yards long, 4 feet wide, and 2 feet deep, in how many days of 12 hours long can 24 men dig a trench 115 yards long, 6 feet wide, and 4 feet deep?

$$\begin{array}{llll} \text{Men} & 24 & : & 240 \\ \text{Length} & 230 & : & 115 \\ \text{Width} & 4 & : & 6 \\ \text{Depth} & 2 & : & 4 \\ \text{Hours} & 12 & : & 10 \end{array} \Bigg\} : : 5 \text{ days.}$$

$$\frac{240 \times 115 \times 6 \times 10 \times 4 \times 5}{24 \times 230 \times 4 \times 2 \times 12} = \frac{5 \times 5 \times 5}{2} = 62\tfrac{1}{2} \text{ days Ans.}$$

4. If a family of 8 persons expend $600 in 9 months, how much will serve a family of 18 persons 16 months?

Ans. $2400.

5. If 120 bushels of oats keep 14 horses 56 days, how many days will 95 bushels feed 6 horses?

Ans. $103\tfrac{4}{9}$ days.

6. If a quantity of provision serves 1500 men 12 weeks, at the rate of 20 ounces a day to a man, how many men will the same provision maintain 40 weeks, at the rate of 8 ounces a day? Ans. 1125 men.

7. If 300 bushels of wheat at $1.25 per bushel will discharge a certain debt, how many bushels at 90 cents a bushel will discharge a debt 3 times as great?

$$\begin{array}{l} .90 : 1.25 \\ 1 : 3 \end{array} \Big\} : : 300 : A.$$

Ans. 1250 bushels.

8. If by travelling 6 hours a day at the rate of $4\tfrac{1}{2}$ miles an hour, a man perform a journey of 540 miles in 20 days, in how many days, travelling 9 hours a day at the rate of $4\tfrac{2}{3}$ miles, will he travel 600 miles? Ans. $14\tfrac{2}{7}$ days.

In this question one of the terms of the dividend, and one of the divisor contain a fraction; to wit, $4\tfrac{1}{2}$ and $4\tfrac{2}{3}$. Reduced to an improper fraction they make $\tfrac{9}{2}$ and $\tfrac{14}{3}$. These fractions may be treated as integers, by merely placing their denominators on the *opposite side of the horizontal line* to that where their respective terms belong.*

9. If a hall 36 feet long by 9 feet wide, would require 36 yards of carpeting 1 yard in width to cover it ; how many yards of the width of 1 ell English, would cover a floor 60 feet long and 27 feet wide ? Ans. 144 yards.

NOTE. In *all* cases of proportion, the terms compared together, must be brought to the *same denomination* ; therefore, in this question, the ell and the yard are reduced to quarters.

10. If 14 men working $8\frac{1}{4}$ hours, can mow 84 acres of grass in 3 days, in how many days, can 12 men working $7\frac{1}{2}$ hours, mow 75 acres ? Ans. $3\frac{17}{56}$† days.

11. If 12 oz. of wool make $1\frac{1}{2}$ yard of cloth $\frac{7}{8}$ wide, how many yards $1\frac{1}{4}$ yard wide, will 16 lb. of wool make ?

$$\left. \begin{array}{c} \frac{3}{4} : 16 \\ 1\frac{1}{4} : \frac{7}{8} \end{array} \right\} : : 1\frac{1}{2}$$

$$\frac{16\times7\times3\times4\times4}{8\times2\times3\times5} = \frac{7\times4\times4}{5} = 22\frac{2}{5} \text{ yards Ans.}$$

In order to bring the pounds and ounces to the same denomination, we reduce the latter to the fraction of 1 lb. 12 oz.$=\frac{3}{4}$ lb.

- After the statement, we arrange the terms whose product is to form the dividend above, and those constituting the divisor below the line, and transfer the denominators of the terms containing fractions to the *opposite side of the line* The whole are then treated as whole numbers.

12. If the transportation of 12 cwt. 2 qr. 8 lb., 206 miles cost $25.75, how far at the same rate, may 3 tons and 3 quarters be carried, for $243 ? Ans. $402\frac{2}{7}$ miles.

The two terms, expressing the quantities, may be brought to hundred weights.

12 cwt. 2 qr. 8 lb.$=12\frac{4}{7}$ cwt.
3 T. 3 qr. $=60\frac{3}{4}$ cwt.

---

* That is, transferring the denominator of the dividend to the divisor, and the denominator of the divisor to the dividend.

† That is, they work 3 days, and $8\frac{13}{24}$ hours the fourth day.

13. If a footman in 12 days, travelling 6 hours à day, perform a journey of 240 miles ; in how many days will he perform one of 720 miles, if he travel 8 hours a day ?
Ans. 27 days.

14. If 20 men in 12 days, working 5 hours a day, can perform a piece of work, how many hours a day, must 15 men work, in order to perform $3\frac{1}{3}$ times as much work in 30 days ?
Ans. $8\frac{8}{9}$ hours.

15. If the transportation of $5\frac{3}{4}$ cwt. 150 miles, cost $24.58, what must be paid for transporting 15 cwt. 1 qr. 22 lb. 64 miles, at the same rate ?
Ans. $28.17+.

16. In what time will $627.50, loaned at 7 per cent., produce as much interest as $2510, at $3\frac{1}{2}$ per cent., will produce in 1 year and 8 months ?
Ans. $3\frac{1}{3}$ years.

17. If a block of marble 2 feet 6 inches long, 1 foot 9 inches broad, and 1 foot 3 inches thick, weigh 9 cwt. 2 qr., what would it weigh, if each of its dimensions were doubled ?
Ans. 3 tons, 16 cwt.

18. 8 workmen laboring 7 hours a day for 15 days, were able to execute $\frac{1}{3}$ of a job on which they were engaged ; in how many days can they complete the residue, by working 9 hours a day, if 4 workmen are added to their number ?
Ans. $15\frac{5}{9}$ days.

19. If 12 men working 9 hours a day for $15\frac{5}{9}$ days, are able to execute $\frac{2}{3}$ of a job, how many men may be withdrawn, and the residue be finished in 15 days more, if the laborers are employed only 7 hours a day ?
Ans. 4 men.

20. If a cistern $8\frac{3}{4}$ feet long, $5\frac{1}{4}$ feet wide, and $6\frac{1}{2}$ feet deep, contain $68\frac{1}{4}$ barrels of water, how many barrels would it hold, if each of its dimensions were doubled ?*
Ans. 546 barrels.

21. If 1000 lb. of wool, would make 752 yards of cloth $1\frac{7}{8}$ yards wide ; what quantity of cloth 2 yards wide, may be made of 1672 lb. of the same wool ?
Ans. $1178\frac{3}{4}$ yards.

---

* This question is susceptible of a much easier method of solution than by compound proportion ; but it would be anticipating another rule to introduce it here.

A NEW SYSTEM OF PROPORTION.

THE common method of statement and solution has been given on the preceding pages, and is perhaps sufficiently plain and accurate for all the purposes of practical business. It conveys, however, but a partial and limited notion of the science of proportion, since it restricts the terms of the statement to a *single* order, and throws the blank invariably in the fourth term.

It is liable, in Compound Proportion, to the further objection, that it requires as many statements as the question contains conditions ; and the ratio of the first two terms of these several statements, is not the same as the ratio of the third to the required term.

The method now to be explained is believed to be more simple, and more conformable to the principles of the science.

The properties of a geometrical proportion consisting of four terms, have been explained on page 92, and the reason assigned why *the product of the means is equal to the product of the extremes.* This equality between their products suggests an obvious method of finding any term of a proportion, which may be blank or unknown : because, if the product of any two numbers be divided by one of the numbers, the quotient will be the other number.

Taking the proportion, 2 : 6=4 : 12, and representing the required term by [ ] we have

$$2 : 6 = 4 : [\ ] - - - - 6 \times 4 = 24 \div 2 = 12 \text{ the 4th term.}$$
$$2 : 6 = [\ ] : 12 - - - - 2 \times 12 = 24 \div 6 = 4 \text{ the 3d }\text{``}$$
$$2 : [\ ] = 4 : 12 - - - - 2 \times 12 = 24 \div 4 = 6 \text{ the 2d }\text{``}$$
$$[\ ] : 6 = 4 : 12 - - - - 6 \times 4 = 24 \div 12 = 2 \text{ the 1st }\text{``}$$

The problem of the Rule of Three is, to find the blank or unknown term of a proportion regularly stated ; and it is obviously indifferent, in which of the four terms the blank falls, if we observe this universal

RULE. If the blank fall in either of the extremes, divide the product of the means by the given extreme ; and if it fall in either of the means, divide the product of the extremes by the given mean.

The terms of a proportion may be variously arranged, without destroying the equality of the ratio of the two couplets. Thus

$$2 : 4 = 3 : 6.$$
$$4 : 2 = 6 : 3$$
$$4 : 6 = 2 : 3.$$

and

$$3 : 6 = 2 : 4.$$
$$6 : 3 = 4 : 2.$$
$$3 : 2 = 6 : 4.$$

Every question in the Rule of Three contains a *supposition* and a *demand :* and the supposition supplies two terms of a proportion and the demand two, one of which is blank or unknown. Now, stating is merely arranging these four terms in pairs, so that the ratio of each of the two couplets shall be the same. For example :—

1. If $ 12 be paid for 8 yards of cloth, what will be the cost of 14 yards of the same cloth? Ans. $ 21.

Here it is evident, that the ratio between the *number* of yards (8) in the supposition, and the number of dollars (12) paid for them, is the same as the ratio between the number of yards (14) in the demand, and the number of dollars to be paid for them ; and inversely. Hence they may be arranged in a proportion accordingly.

yd. $ yd. $ $ yd. $ yd.
8 : 12 = 14 : [ ];   Or, 12 : 8 = [ ] : 14.

Again, the ratio of the number of yards in the supposition, to the number of yards in the demand, is the same as the ratio of the price of the first to the price of the last ; and inversely. And we may say

yd.   yd. $ $ yd.   yd. $ $
8 : 14 = 12 : [ ];   Or, 14 : 8 = [ ] : 12.

In all the foregoing statements, the product of 14 and 12 is by the rule given above, to be divided by 8, and the same result is produced.

But we may invert the four proportions above, so as to bring the blank on the left of the sign of equality, and thus produce four other forms of statement.

Perhaps the simplest and most intelligible method for learners is, to form one couplet of the proportion out of the two terms supplied by the supposition, and the other couplet out of the two contained in the demand ; observing always to *arrange them in the same order.* It is indifferent with which branch of the question we begin.

2. What is the height of a steeple, which casts a shadow 198 feet on level ground, if a perpendicular staff 4 feet long, cast a shadow at the same time 7 feet long ?

h.   s.   h.   s.   s.   h.   s.   h.
[ ] : 198 = 4 : 7;   Or, 198 : [ ] = 7 : 4.

In the first statement the order of the two terms taken from the demand is *height, shadow* ; and the order of the two terms taken from the supposition is correspondent, *height shadow.* In the second, the order is reversed : it is *shadow, height,* in both couplets.

3. If $\frac{2}{5}$ of an ell English cost £$\frac{1}{3}$, what will 12$\frac{1}{2}$ yards of cloth cost ?

$$\begin{array}{cccccccc} \text{c.} & £ & \text{c.} & £ & & £ & \text{c.} & £ & \text{c.} \\ \frac{2}{5} &:& \frac{1}{3} &=& 12\frac{1}{2} &:& [\ ]. & \text{Or,} & \frac{1}{3} :& \frac{2}{5} &=& [\ ] :& 12\frac{1}{2}. \end{array}$$

Observe, that the corresponding terms of the statement, $\frac{2}{5}$ and 12$\frac{1}{2}$, are not in the same denomination : one is *yards,* and the other is *ells.* Before multiplication, they are not reduced to quarters. In proportion the rule is universal, that *like terms must be reduced to the same denomination.*

When fractions occur in any of the terms of the statement, the denominators of the extremes are multiplied into the numerators of the means, and the denominators of the means into the numerators of the extremes. If there are mixed numbers, they must be reduced to improper fractions.

For convenience, place the terms whose product is to form the dividend above a line, and the divisor beneath, and if there are fractions, place the denominators of the dividend with the divisor, and the denominators of the divisor with the dividend.* Thus, in the last example, 12$\frac{1}{2}$ $(=\frac{25}{2})$ and $\frac{1}{3}$ are the dividend ; and their denominators 2 and 3, are placed with the divisor $\frac{2}{5}$, and the denominator of $\frac{2}{5}$ with the dividend. Moreover, 4 is placed in the dividend to reduce yards to quarters, and 5 with the divisor to reduce $\frac{2}{5}$ of an ell to the same denomination.

$$\frac{\text{Dividend}}{\text{Divisor}}\ \ \frac{25\times 1\times 5\times 4}{2\times 3\times 3\times 5}=£\ \tfrac{50}{9},\ \text{or}\ £5,\ 11\ \text{s.}\ 1\tfrac{1}{3}\ \text{d.}$$

By this method, vulgar fractions present no more difficulty than integers, and in fact, are treated precisely like them.

4. What will 7$\frac{7}{9}$ gallons of molasses come to, if 8$\frac{3}{4}$ gallons cost 4$\frac{1}{5}$ dollars ?

$$\begin{array}{cccccccc} \text{gal.} & \$ & & \text{gal.} & \$ \\ 7\frac{7}{9} &:& [\ ] &=& 8\frac{3}{4} :& 4\frac{1}{5}; & \text{or} & \frac{70}{9} :& [\ ] &=& \frac{35}{4} :& \frac{21}{5}. \end{array}$$

It is more convenient to use a *perpendicular* line ; placing the dividend on the right, and the divisor on the left ; as we may then omit the signs of multiplication.

Any two numbers on opposite sides of the line, that are divisible by the same number, may be reduced by the method shown on page 31 ; and their quotients substituted for them ; a pencil b ing drawn across the divided numbers, to show that they are disposed of. Thus in the last example.

$$\begin{array}{c|c} 2 & 25 \\ 3 & 1 \\ 3 & 5 \\ 5 & 4\ \ 2 \\ \hline 3\times 3 & 25\times 2 \\ 9\,)\,50\,(\,5\frac{5}{9}. \end{array}$$

$$\frac{\text{Dividend}}{\text{Divisor}} \frac{70 \times 21 \times 4}{9 \times 5 \times 35} = \frac{2 \times 7 \times 4}{3 \times 5 \times 1} = \frac{56}{15}, \text{ or } \$3\frac{11}{15}. \text{ Ans.}$$

5. If 7 lbs. of sugar cost 75 cents, how many pounds can be bought for 9 dollars ?

s. s. s.
.75 : 7 = 9 : [ ].

$$\frac{7 \times 9.00}{.75} = 84 \text{ lbs. Ans.}$$

The pupil may solve by this method any of the questions found in this book, under Simple Proportion.

The method of statement above explained may be applied with equal facility, and still greater convenience, to

COMPOUND PROPORTION.

Here, as in the Single Rule of Three, every question consists of two branches, the supposition and the demand, each of which supplies two leading terms of a proportion. It differs from the latter only by connecting with these leading terms certain conditions, or modifying incidents, which influence the result.

6. If $ 100 capital produce $ 7 profit, how much profit will $ 480 produce ? is a question in Simple Proportion.

But if we connect with the $ 100 the condition, that it be invested for 12 months, and with the $ 480 the condition, that it be invested for 30 months, it becomes a question in Compound Proportion, and these conditions must be included in the statement: that is, they must be connected by the sign of multiplication with the term which each modifies. The statement is

c. p. c. p.
100 × 12 : 7 = 480 × 30 : [ ].

The reason of multiplying the leading terms by their modifying conditions, will be apparent if we consider that,

$ 100 for 12 months = $ 1200 for 1 month, and
$ 480 for 30 months = $ 14400 for 1 month.

The effect, therefore, of the multiplication in this statement is, *to equalize the time*, during which the capital in the two branches of the question is productive, so that the results may be proportional. The statement above is equivalent to the following :

c. p. c. p.
1200 : 7 = 14400 : [ ]. Ans. $ 84.

7. If 5 men mow 12 acres of grass in a given time, how many men will mow 36 acres in the same time ? is a ques-

$$\text{m.} \quad \text{a.} \quad \text{m.} \quad \text{a.}$$
$$5 : 12 = [\ \ ] : 36.$$

But let the question be : If 5 men in 2 days, working 6 hours a day mow 12 acres ; how many men working 3 days, 10 hours a day will mow 36 acres? and we have two conditions to connect with each set of men. The statement is :

$$\text{m.} \qquad \text{a.} \quad \text{m.} \qquad \text{a.}$$
$$5 \times 2 \times 6 : 12 = [\ \ ]\ 3 \times 10 : 36.$$

$$\frac{5 \times 2 \times 6 \times 36}{12 \times 3 \times 10} = \frac{6}{1} \qquad \text{Ans. 6 men.}$$

There is very little difficulty in distinguishing the leading terms of the proportion from their incidents ; or in knowing to which of the terms the incidents belong. In each branch of the question, the leading terms have in general the relation to each other of *cause* and *effect ;* or the grammatical relation of *agent* and *object ;* and the incidents commonly denote time, rate, dimension, &c.

8. How many men will dig a trench 135 yards long and 4 yds. wide in 8 days, if 16 men will dig one 54 yards long, and 5 yds. wide in 6 days ?

In this question, both the agent and object, (that is the men and the trench,) have modifying conditions connected with them. The men are to work a specified number of days, and each trench has stated dimensions.

$$\text{m.} \quad \text{t.} \qquad \text{m.} \qquad \text{t.}$$
$$[\ \ ]\ 8 : 135 \times 4 = 16 \times 6 : 54 \times 5. \qquad \text{Ans. 24 men.}$$

We omit in this statement, the term (1) denoting the trench, because, being unity, it does not at all affect the result. Moreover, we may always omit in stating, any term *which is the same in both branches of the question,* unless it be necessary as one of the constituent terms of a regular proportion.

9. If 20 horses eat 70 bushels of oats in 3 weeks, how many bushels will 6 horses eat in the same time ?

$$\text{h.} \quad \text{b.} \quad \text{h.} \qquad \text{b.}$$
$$20 \times 3 : 70 = 6 \times 3 : [\ \ ]. \qquad \text{Ans. 21 bushels.}$$

This question properly belongs to Compound Proportion ; but the condition 3 being the same in the means and in the extremes may be omitted in the statement.

10. If 6 men in 12 twelve days build a wall, in how many days would 20 men build it ?

Questions of this kind are usually classed in the "Single Rule of Three Inverse." They in fact, however, belong to the Double Rule of Three. But one of the terms (1 wall) being the same in the supposition and in the demand, does not influence the result. It may be stated

m.   w.   m.   w.

$6 \times 12 : 1 = 20$ [1 : 1.   Ans. $3\frac{2}{3}$ days.

All the questions usually considered as belonging to In- verse Proportion, admit of similar statement and solution.

11. How many yards of cloth $\frac{3}{4}$ yd. wide are equal to 30 yards $1\frac{1}{4}$ yd. wide ?

y.   q.   y.   q.

[ ] $\frac{3}{4} : 1 = 30 \times 1\frac{1}{4} : 1$ Ans. 50 yds.

The 1 in both branches of this statement, represents the whole quanti- ty of each kind. It might have been equally well represented by any other number, as the only object is to have the requisite number of terms to constitute a proportion. The foregoing explanations may be summed up in the following

RULE. State the question by arranging the leading terms of the supposition and of the demand, in the same order ; con- necting with each term by the sign of multiplication, the con- ditions which influence, or properly belong to it. Then, if the blank fall in either of the extremes, divide the product of the means by the product of the extremes ; and if it fall in either of the means, divide the product of the ex- tremes by the product of the means.

NOTE. The corresponding terms and conditions must in all cases be reduced to the same denomination previous to multiplication and division.

12. If 25 men working 10 hours a day for 9 days, dig a ditch, 36 yds. long, 12 feet broad, and 6 deep ; how many hours a day, must 15 men work, in order to dig a ditch 48 yds. long, 8 feet broad, and 5 ft. deep in 12 days ?

m.   d.   m.   d.

$25 \times 10 \times 9 : 36 \times 12 \times 6 = 15$ [ ] $12 : 48 \times 8 \times 5$.

Ans. $9\frac{7}{27}$ hours.

13 How many acres of grass will 15 men mow in $3\frac{3}{4}$ days, by working 9 hours a day, when 4 men in $2\frac{1}{2}$ days mow $6\frac{2}{3}$ acres by working $8\frac{1}{4}$ hours a day ?

a.   m.   a.   m.

[ ] $: 15 \times 3\frac{3}{4} \times 9 = 6\frac{2}{3} : 4 \times 2\frac{1}{2} \times 8\frac{1}{4}$

$3\frac{3}{4} = \frac{15}{4}.$ $6\frac{2}{3} = \frac{20}{3}.$ $2\frac{1}{2} = \frac{5}{2}.$ $8\frac{1}{4} = \frac{33}{4}.$

$$\frac{\text{Dividend}}{\text{Divisor}}\ \frac{15 \times 15 \times 9 \times 20 \times 2 \times 4}{4 \times 4 \times 5 \times 3 \times 33}\qquad \text{Ans. } 40\frac{10}{11} \text{ acres.}$$

14. If 12 oz. of wool make $2\frac{1}{2}$ yards of cloth $1\frac{1}{2}$ yd. wide ; how many pounds of wool is required to make 150 yards 13 yd. wide ?

                 w.   c.   w.   c.

12 oz.$=\frac{3}{4}$ lb. $\frac{3}{4}$: $2\frac{1}{2}\times1\frac{1}{2}$ = [ ] : $150\times1\frac{3}{4}$.

                                        Ans. $52\frac{1}{2}$ lb

15. In what time will $ 36 at 7 per cent. per annum, gain $ 3.78 ?

     p.   i.   p.   i.

     100×12 : 7 = 36 [ ] : 3.78.   Ans. 18 months

16. In what time will $ 75 at 5 per cent. per annum. produce as much interest as $ 450 at $7\frac{1}{2}$ per cent. will produce in $3\frac{3}{5}$ years ?

     75 [ ] 5 : 1 = $450\times3\frac{3}{5}\times7\frac{1}{2}$: 1.   Ans. $32\frac{2}{5}$ years.

17. If the transportation of 1 ton 25 miles be $ 6, what ought to be paid for transporting 17 cwt. 3 qrs. 45 miles ?

     w. $ w. $

     20×25 : 6 = $17\frac{3}{4}\times45$: [ ].   Ans. $ $9.58\frac{1}{2}$.

The 1 ton is reduced to cwt. because the corresponding term in the demand is cwt.

18. A wall was to be built 700 yards long, in 29 days, and 12 men being employed on it for 11 days, completed only 220 yards. How many men must be added to the number to finish the wall in the specified time ?

     700$-$220$=$480.   29$-$11$=$18.

         12×11 : 220 = [ ] 18 : 480.   Ans. 4 men.

It is unnecessary to multiply examples in illustration of the system now proposed. This method of operation, it is believed, will be found preferable to the one commonly practised. It conforms to, and preserves the true principles of a proportion, in making it consist of four terms. It admits of all the transposition in the order of the terms consistent with the nature of a proportion, without affecting the equality of their ratios or the result of the solution. It is very questionable, whether scholars taught upon the common system, even suspect that there is more than one method of stating; or have any very correct idea of the true nature of a proportion. They would be apt to conclude, that the blank term of a statement must necessarily be the *fourth*. It has already been seen, that it may fall in any of the four terms indifferently, but it may not be amiss to show by an example, in how many different ways a question in the Rule of Three may be stated, in strict conformity with the truth of the proportion, and without varying the result.

If $1\frac{3}{4}$ yards of gingham cost 54 cents what will 87 yards cost ?

     y. $ y. $ y. $ y. $

     $1\frac{3}{4}$: .54 = 87 : [ ].   87 : [ ] = $1\frac{3}{4}$: .54.

     $ y. $ y. $ y. $ y.

     .54 : $1\frac{3}{4}$ = [ ] : 87.   [ ] : 87 = .54 : $1\frac{3}{4}$.

     y.   y. $ $ $ $ y.   y.

     $1\frac{3}{4}$: 87 = 54 : [ ]   54 : [ ] = $1\frac{3}{4}$: 87

QUESTIONS.

What is meant by ratio ?

How do we express the ratio of two numbers ?

Does it make any difference which of the numbers is taken for the numerator, and which for the denominator ?

Is the ratio of two numbers the same as the quotient resulting from the division of one of them by the other ?

What name is given to the first, and what to the second term of a couplet ?

When two pairs of numbers have the same ratio, what do they constitute ?

What name is given to the two outer terms, and what to the two middle terms of a proportion ?

What relation is there between the product of the means, and the product of the extremes of a proportion ?

Why is the product of the means equal to the product of the extremes ?

What rule is founded on this equality ?

Explain the common method of stating in the Single Rule of Three.

How do we proceed after a question is stated ?

Why must the two like terms be brought to the same denomination ?

After the division, if there be a remainder, it may be treated in three different methods ; what are these methods ?

What is compound proportion ?

Which of the given terms is placed for the last term in the statement ?

Which terms are multiplied together for the dividend, and which for the divisor ?

If any of the terms of the dividend are fractions or mixed numbers, on which side of the horizontal line are their denominators to be placed ?

And if the terms of the divisor contain fractions, where are their denominators set ?

In stating according to the method laid down in the New System of Proportion, how do you arrange the terms ?

With which of the leading terms of the proportion must the conditions be connected ?

If a term or a condition is the same in the supposition and in the demand, need it be inserted in the statement ?

Do the questions commonly considered as involving "Inverse Proportion" belong to Simple or Compound Proportion ?

If the bracket denoting the required number, fall in one of the extremes, which terms are to be multiplied together for the dividend ?

If it fall in either of the means, which terms constitute the dividend?

INTEREST.

INTEREST is a premium paid for the use of money. It is computed at a certain per cent. per annum. *Per cent.* means by the hundred ; and interest is so many dollars on a hundred. The sum on which it is paid is called the *Principal*, and the per cent., the *Rate.* The principal and interest added together are called *Amount.*

The rate of interest is fixed by law, and varies in the different states. In New-York, it is **7** per cent., but in New-England, 6 per cent.

The taking of more than the legal rate, is called *usury ;* and an agreement to do so, in some of the states, subjects the creditor to forfeiture of the whole debt.

For computing the interest on any sum for 1 year.

RULE. Multiply the principal by the rate per cent., and divide the product by 100.

1. What is the interest of £139 11s. 8½d. for 1 year, at 6 per cent. ?

```
    139   11   8½
                6
  ──────────────────
  8|37   10   3
    20
  ──────
  7|50
    12
  ──────
  6|03
     4
  ──────
   |12     Ans: £8   7s.   6d.
```

Dividing by 100 is merely cutting off the two right hand figures by a line. The £37 remaining are reduced to shillings, and the 10s. added in ; and the product again divided by 100. The next remainder is treated in the same manner.

Every question in interest, is in reality a question under the Single Rule of Three, and admits of statement.

Thus the first example may be stated :

$$\pounds \quad \pounds \quad \text{s.} \quad \text{d.} \quad \pounds$$
$$100 : 139 \quad 11 \quad 8\tfrac{1}{2} : : 6 : \text{A.}$$

But as the first term is constant, a formal statement is unnecessary.

2. Find the interest of £450 for a year, at 5 per cent. per annum.   Ans. £22 10s.

3. Find the interest of £715 12s. 6d. for a year, at 4½ per cent. per annum.   Ans. £32 4s. ¾d.

4. Find the interest of £230 10s. for a year, at 4 per cent. per annum.   Ans. £9 4s. 4¾d.

Before multiplying by the rate per cent., we may reduce the shillings, pence, &c., to the decimal of a pound. Thus in the last example :

10s.=.5 of £1.

```
        .  230.5
               4
          _____
          9|22.0
            20
          _____
          4|40
            12
          _____
          4|80
             4
          _____
          3|20      Ans. £9  4s. 4¾d.
```

FEDERAL MONEY.

The general rule is applicable as well to Federal as Sterling money. Dividing by a hundred *decimally*, it will be recollected, is simply removing the decimal point two places toward the left. The operation, therefore, is a very simple one.

5. Required the interest of $ 135.25 for 1 year, at 6 per cent.

```
        135.25
             6
        _____
```

Ans. $ 8.1150

The decimal point for multiplication, falls between the 1 and 5 Removing it two places toward the left for division by 100, gives the interest $ 8.11½.

6. What is the interest of the same sum at 7 per cent.

```
        135.25
             7
        _____
```

Ans. $ 9.4675

7. What is the interest of $19.52 for 1 year, at 5½ per cent. ?

. 19.52
5½
_____

9760
973
_____

Ans. $ 1.0736

If the interest be required for a number of years, the rate per annum may be multiplied by the given number of years ; or, the interest for 1 year may be multiplied by the given number of years.

8. Required the interest of $ 85.45 for 5 years, at 6 per cent.

6×5=30 per cent.   85.45
30
_____

Ans. $ 25 6350

9. The interest of the same sum at 7 per cent.

7×5=35 per cent.   85.45
35
_____

42725
25635
_____

Ans. $ 29.9075

If the interest be required for parts of a year, correspondent parts of the interest for 1 year may be taken ; or, we may take parts of the rate, and multiply by them.

10. What is the interest of $ 164.20 for 1 year and 4 months, at 7 per cent. ?

164.20
7
_____

4 m. ⅓ of 1 year.   ⅓)11.4940
3.831
_____

Ans. $15.325

The interest of the same sum at 6 per cent., may be more conveniently found by the second method.

Rate for 12 mo.=6
" for 4 " $=\frac{1}{3}$ of 6=2 6+2=8 per cent.

```
  164.20
       8
  ------
Ans. $ 13.1360
```

Observe, that at 6 per cent. for 12 months, the rate is equal to just half the number of months. At this rate, therefore, we may always find the interest, by multiplying by half the number of months.

11. Required the interest of $ 198 for 2 years and 2 months, at 6 per cent.

```
                          198
2 y. 2 mo.=26  ½=13        13
                          ----
                          594
                         198
                         ----
                  Ans. $ 25.74
```

12. Required the interest of $ 38.16 for 5 months at 6 per cent.

```
                          38.16
½ of 5 mo.=2½.              2½
                          -----
                          7632
                          1908
                          -----
                 Ans. $ .9540
```

7 per cent. equals $\frac{7}{6}$ of 6 per cent. Therefore, by adding $\frac{1}{6}$ to the interest of 6 per cent., we have the interest at 7 per cent.

```
    ⅙|.954
      .159
     -----
Interest of the same
sum at 7 per cent.=   $1.113
```

13. Required the interest of $ 64.58 for 3 years, 5 months, and 10 days, at 7 per cent. Ans. $ 15.57.

4 months $=\frac{1}{3}$ of 1 year.
1 month $=\frac{1}{4}$ of 4 months.
10 days $=\frac{1}{3}$ of 1 month.

In the computation of interest, 30 days are reckoned a month, 60 days, 2 months.

At 6 per cent. for 12 months, the rate for 2 months will be $\frac{1}{6}$ of 6 per cent.$=1$ per cent.   Therefore the rate

for			for		
1 month$=\frac{1}{2}$ or .5 per cent.			6 days$=\frac{1}{10}$ or .1 per cent.		
20 days	$\frac{1}{3}$	" "	12 "	.2	" "
15 "	$\frac{1}{4}$	" "	18 "	.3	" "
10 "	$\frac{1}{6}$	" "	24 "	.4	" "

To compute interest at 6 per cent, for any number of days above specified, we have but to apply the general rule, and for intermediate numbers, take aliquot parts.

14. Required the interest of $ 112 at 6 per cent. for 6 days.   Ans. $ .112 ; or 11 cents, 2 mills.

Multiplying by .1, and dividing by 100, is simply removing the decimal point 3 places toward the left.

15. Required the interest of $ 112 for 9 days.

Interest for 6 days - - - .112
       3 days$=\frac{1}{2}$ - - - - .056
                                   ———
                          Ans. $ .168

16. What is the interest of $ 234. for 5 days, at 6 per cent.?
Interest for 6 days   .234
       Subtract $\frac{1}{6}$ 39
                                   ———
                          Ans. $ .195

On account of the facility of calculating interest for days at 6 per cent., it is generally best to do so, and then add or subtract, according as the rate is more or less than 6 per cent.

17. Required the interest of $ 158 for 2 months and 8 days, at 7 per cent. ?

At 6 per cent. interest for 2 months   -   1.58
       "   "   6 days - -   .158
       "   "   2 "   - -   .052
                                   ———
                                  · 1.790
              Add $\frac{1}{6}$ .298
                                   ———
                          Ans. $ 2.088

The interest at 5 per cent. is found by subtracting $\frac{1}{6}$.

18. Required the interest of $ 789 for 2 years, 3 months, and 24 days, at 7 per cent. Ans. $ 127.95.

19. Required the interest of $ 37.50 for 4 years 11 months, and 18 days, at 7 per cent. Ans. $ 13.037.

The reckoning of 30 days to the month is authorized by custom, and is sufficiently accurate for ordinary purposes. As, however, a year is not 360 (30×12,) but 365 days, if perfect exactness be desirable, the interest may be computed for a year, and the result taken for the third term of a statement in the Rule of Three, the given number of days being made the second, and 365 the first term.

20. What is the interest of $ 800 for 1 year, 3 months, at 6 per cent.? Ans. $ 60.

21. Find the interest of $ 58.11 for 1 year and 11 months, at 6 per cent. Ans. $ 6.68.

22. Find the interest of $ 273.51 for 2 years 20 days, at 7 per cent. Ans. $ 39.354.

23. Find the interest of $ 650.82 for 3 years, 4 months, at 5 per cent. Ans. $ 108.47.

24. Find the interest of £75 8s. 4d. for 4 years 7 months, at 6 per cent. Ans. £20 14s. 9$\frac{1}{2}$d.

25. Find the amount of $ 279.87 for 2$\frac{1}{2}$ years, at 7 per cent. Ans. $ 328.84.

NOTE. Amount is the principal and interest added together.

26. Find the amount of $ 683.20 for 5$\frac{3}{4}$ years, at 7$\frac{1}{2}$ per cent. per annum. Ans. $ 977.83.

27. Find the amount of $ 17.34 for 6 years, 5 months, and 18 days, at 6 per cent. per annum. Ans. $ 24.067.

28. Find the interest on $ 8.50 for 1 year, 7 months, at 6 per cent. Ans. $ .807.

29. Find the interest on $ 73 for 10 months, at 6 per cent. Ans. $ 3.65.

30. Find the interest on $ 675 for 1 month, 21 days, at 7 per cent. Ans. $ 6.69.

31. Find the interest on $ 467.17 for 3 years, and 5 months, at 7 per cent. per annum. Ans. $ 111.73.

32. Find the amount of £320 8s. for 10 years, at 6 per cent. per annum.   Ans. £512 8s. 9¼d.

PARTIAL PAYMENTS.

It is customary, when payments in part are made on a note, bond, &c., to write the sum paid on the *back* of the instrument, (from which circumstance it is called *indorsement ;*) or to give a receipt specifying that it is to be applied in payment of such instrument.   The rule for computing interest in such cases, adopted by most of the individual states, and by the Supreme Court of the United States, is laid down in a decision of Chancellor Kent, of the state of New-York.   According to this decision, the method of casting interest is, to " *apply the payment in the first place to the discharge of the interest then due.   If the payment exceeds the interest, the surplus goes towards discharging the principal, and the subsequent interest is to be computed on the balance of principal remaining due.   If the payment be less than the interest, the surplus of interest must not be taken to augment the principal ; but interest continues on the former principal, until the period when the payments taken together exceed the interest due, and then the surplus is to be applied toward discharging the principal ; and interest is to be computed on the balance, as aforesaid.*"

. RULE. Compute the interest to the date of the first payment which, by itself, or with the addition of a previous payment, or payments, *exceeds* the interest then due ; add the interest to the principal, and subtract the payment (or payments.) The remainder forms a new principal, with which proceed as before, to the time of final settlement.

33. A note for $ 2000 was given January 4, 1827, on which there were the following payments :

February 19, 1828 - -	$ 400,
June 29, 1829 - - -	$ 1000,
November 14, 1829 -	$ 520.

How much remained due December 24, 1830, interest at 6 per cent. ?

Pr̄ucipal - - - $ 2000
Interest from Jan. 4 to Feb. 19, (13½ mo.) -   135

First Amount - . -   2135
First payment, - - - - - - - -   400

Balance forming a new principal - - - -   1735
Interest from February 19 to June 29, (16⅓ m.) 141.69

Second Amount - -   1876.69
Second payment - - - - - - - -   1000.

Balance forming a new principal - - - -   876.69
Interest from June 29 to November 14, (4½ m.) 19.72

Third Amount - - -   896.41
Third payment - - - - - - - -   520.

Balance forming a new principal - - -   376.41
Interest from Nov. 14 to Dec. 24, (13½ m.) -   25.40

Balance due on taking up the note - - - $ 401.81

34. A note was given February 1, 1830, for the payment of $ 500, on which there were indorsements as follows: May 1, 1830, $ 40 ; November 14, 1830, $ 8 ; April 1, 1831, $ 12 ; May 1, 1832, $ 60.

What is the balance due on the note September 16, 1832 ; interest at 7 per cent. ?

Principal - - - - $ 500
Interest to May 1, 1830, (3 m.) - - - -   8.75

First Amount - - -   508.75
First payment - - - - - - - -   40.

Balance forming a new principal - - - -   468.75
Interest to May 1, 1832, (2 years) - - -   65.62

$$\left.\begin{array}{l} \$\ 8 \\ \$\ 12 \\ \$\ 60 \end{array}\right\} = \$\ 80 \text{ a sum exceeding the interest due}$$

Second Amount - - -	534.37
= $80 a sum exceeding the interest due May 1, 1832.	80.
Balance forming a new principal - - - -	454.37
Interest to September 16, 1832, (4½ m.) - -	11.92
Balance due on taking up the note, - - -	$466.29

NOTE. The second and third indorsements, not being singly, or together, equal to the interest due at the time they were paid, it is computed from the first payment to the fourth, which, in conjunction with the two previous payments, exceeds the interest then due.

35. A note of hand dated April 4, 1832, was given for the payment of six hundred dollars, on which there were indorsements as follows : July 10, 1832, $84.60 ; November 22, 1832, $10 ; April 30, 1833, $14 ; December 5, 1833, $309. What was the balance due on taking up the note, April 5, 1834.   Ans. $251.03.

$864.   *Albany, July* 10, 1830.

Four years from date, we jointly and severally promise, for value received, to pay to the order of George R. Guernsey, eight hundred and sixty-four dollars, with interest.

ROBERT C. DUNCAN,
DAVID JOHNSTON.

On this note are indorsements as follows :

 April 6, 1831 - - - - $34
 June 21, 1832 - - - - 300
 February 26, 1833 - - 180
 January 1, 1834 - - - 40

What was the balance due at the maturity of the note ?

COMPOUND INTEREST.

When interest payable at stated periods is forborne ; or a debt remains unpaid after it falls due, it is equitable to require interest upon interest.

RULE. Compute the interest up to the time that it became payable, and add it to the principal; then cast the interest on that amount for the next period, and add it to *its* principal, and so on. The first principal subtracted from the last amount, will give the compound interest for the whole time.

37. What is the amount of $1000 for 3 years, compound interest, at 7 per cent., payable annually?

```
Principal  - - - - - - $1000
Interest for 1 year - - -      70
                              ─────
Principal for second year -  1070
Interest for second year  -    70.49
                              ─────
Principal for third year  -  1140.49
Interest for third year - -    79.83
                              ─────
Amount for 3 years - - - $1220.32 Ans.
Deduct the first principal -  1000
                              ─────
Compound interest  - - - $220.32
```

38. What is the compound interest of $750, for 4 years, at 6 per cent.?   Ans. $196.85¼.

39. Find the amount of $876.90 for 3½ years, at 6 per cent.; interest to be paid annually.   Ans. $1075.73.

When interest is to be paid semi-annually, it is to be computed for periods of half a year, and added to the principal, as in the foregoing examples.

PROBLEM I. The amount, principal, and time given, to find the rate.

40. At what per cent. will $950.75, amount to $1235.975 in 5 years?

This problem may be solved by a statement in compound proportion, thus:

$$950.75 \; : \; 100 \atop 5 \; : \; 1 \Big\} \; 285.225$$

This statement may be reduced to the following

RULE. Multiply the interest by 100, and multiply the principal by the given number of years ; then divide the former product by the latter.

NOTE. The interest is obtained by subtracting the principal from the amount.

$285.225 \times 100 = 28522.5$
$950.75 \times 5 = 4753.75) 28522.50 (6$ per cent. Ans.
               28522 50

41. At what rate per cent., will $ 340.25 amount to $ 626.06 in 12 years ?   Ans. 7 per cent.
42. At what rate will $ 324.61 amount to $ 430.108 in 5 years, and 5 months ?   Ans. 6 per cent.

PROBLEM II. The amount, principal, and rate given, to find the time.

43. In what time will $ 340.25 amount to $ 626.06 at 7 per cent. ?

Amount - - 626.06
Principal - 340.25·
         ————— $340.25 : 100$ $\left.\begin{array}{c} \end{array}\right\}$: : $\begin{array}{c} y. \\ 1 \end{array}$
Interest - - 285.81 $7 : 285.81$

This statement is the basis of the following

RULE. Multiply the given interest by 100, and divide the product by the product of the principal and rate.

44. In what time will $ 730 amount to $ 975.99, at 6 per cent. ?   Ans. 5 years, 7 m. 12 d.
45. In what time will any sum, say $ 500, double, at 7 per cent. per annum ?   Ans. $14\frac{2}{7}$ years.
46. In what time will any sum double at simple interest, 6 per cent. per annum ?   Ans, $16\frac{2}{3}$ years.

Regarding Simple Interest as a branch of the Rule of Three, in which one of the terms is *constant*, we have preferred the method of multiplying by the rate per cent., and dividing by 100. But, by a principle already explained, we may divide the *rate* by 100, and multiply by the quotient, and obtain the same result. .In this case, the *multi-plier*, at any given rate will be constant :

At 6 per cent. $\frac{6}{100}$=.06

At 7 per cent. $\frac{7}{100}$=.07 ; and so of other rates.

Multiplying the principal by these decimals, we obtain precisely the same result as by the general rule for casting interest for a year at these respective rates. This method may, if the pupil choose, be employed in calculating Commission ,Insurance, &c.

COMMISSION.

Commission is a certain per centage paid to a person, for his services in buying and selling for his employer.

1. If a broker negotiate a loan of $ 5425 for his employer, what is his commission, at 2 per cent. ?

2 per cent.=.02.

$$\begin{array}{r} 5425 \\ .02 \\ \hline \end{array}$$

Ans. $ 108.50

2. What is the commission on $ 3568 at 1¾ per cent. ?

Ans. $ 62.44.

3. What is the commission on £843 10s. at 1¼ per cent.?

Ans. £10 10s. 10¼d.

4. If a man sell goods on commission, to the amount of $ 12450.75 ; what will be his compensation at 5 per cent., and how much will remain for his employer ?

He receives $ 622.53,

His employer $ 11828.21.

5. If a broker sell stocks to the amount of $ 10000 ; what is his commission at ¾ of 1 per cent. ?

Ans. $ 75.

INSURANCE.

Insurance is a contract to make good losses or damage, which may accrue to ships, buildings, or goods, from perils of the sea, from fire, or other accident.

For this security, the owner pays a premium of a certain per cent., on the value of the property insured.

The written instrument or contract of insurance, is called a *policy*.

6. What is the premium for insuring a dwelling house, valued at $2875, against loss, or damage by fire, at $\frac{2}{8}$ of 1 per cent. ?   Ans. $10.78.

7. If an insurance of $25000 be taken on a ship and cargo, returning from Canton to New-York ; what is the premium at $4\frac{1}{2}$ per cent. ?   Ans. $1125.

8. If a stock of goods be insured for $4125 at $\frac{3}{4}$ of 1 per cent., what is the premium ?   Ans. $30.93+.

STOCKS.

Stock is the name given to the capital of banking and other incorporated companies, or to funds established by government. It consists of shares, commonly of $50 or $100. Stocks are transferable from one person to another, and may be bought and sold like other property.

When the shares will fetch in market their nominal value, they are said to be *at par ;* when they sell for more than their nominal value, they are *at an advance,* or *above par ;* when for less, they are said to be at a discount or *below par.* The advance or discount upon the par value of stocks is stated at so much *per cent.*

9. Sold 15 shares of bank stock at $5\frac{1}{2}$ per cent. advance, their par value being $100 a share. How much did I receive for them ?

$$1500$$
$$5\frac{1}{2}$$

$$7500$$
$$750$$

$$\$ 82.50$$

$$100 \times 15 = 1500$$
$$82\frac{1}{2}$$

Ans. 1582\frac{1}{2}$.

10. Bought 25 shares of railroad stock, at $5\frac{1}{4}$ per cent. discount; nominal value $50 a share. How much did they cost me ?   Ans. $1185.94.

11. If I buy 75 shares of the stock of a newly incorpo-

rated bank, and sell them at $9\frac{3}{8}$ per cent. advance, what will be my profit, the shares having cost me $ 50 each ?

Ans. 351.56.

12. When a share of the United States Bank stock sold for $ 112½, the nominal value being $ 100, what were $ 2000 of that stock worth? Ans. $ 2250.

Here the advance is 12½ per cent.

13. If $ 100 of stock in an Insurance Company, sell for $ 96⅝, what are $ 1200 of the stock worth ?

The stock is $3\frac{3}{8}$ per cent. below par. Ans. $ 1159.50.

DISCOUNT.

DISCOUNT is applicable only to demands not drawing interest, and to notes on which the interest is paid in advance, whereon the drawer receives a sum, which, at the customary rate of interest, will amount to the face of the note in the specified time.

The problem then of Discount is, from the rate, time, and amount, to find the principal.

This is solved by the Rule of Three. We take any sum—for convenience, 1 dollar, or 1 pound—and find its amount at the given rate, and time ; and then say, as this amount is to the given sum, (which is also *amount,*) so is 1 dollar—to the principal or present worth of the sum in question.

1. For example ; a note for $ 246 is payable 2 years hence, and not on interest. If it be paid now, what is its present worth ?

The interest of $ 1 at 7 per cent. is .07, and for 2 years .14. Adding this interest, we have the amount $ 1.14. The statement would be ;

Amount. Amount. Prin.

1.14 : 246 : : 1 : A.

But the third term is constant, being always unity. We may therefore omit it.

RULE. Divide the given sum by the amount of $1 for the given time and rate.

```
1.14)246.00(215.78+ Ans.
     228
     ───
     180
     114
     ───
     660
     570
     ───
     900
     798
     ───
     1020
      912
     ───
```

Subtracting the present worth from the given sum, we have the discount.

```
246
215.78
──────
```
Discount $30.22

If we take the present worth or principal, and cast the interest for the given time, we shall obtain for amount the original sum.

2. If a note for $925 be payable without interest, 1 year 8 months hence; what is its present worth at 6 per cent.? Ans. $840.91.

3. If I buy goods in Montreal, to the amount of £615 15s. on a credit of 7 months; how much ought to be deducted, if I pay down, discount being 4½ per cent. per annum? Ans. £15 15s.

4. What is the present worth of $756, one half payable in 6 months, and the other half in a year, discount at 7 per cent.? Ans. $718.49.

5. What is the discount on $600, payable in 4 years, at 5 per cent. per annum? Ans. $100.

6. What is the present worth of a legacy of $1200 to be paid when the legatee comes of age, he being 16 years old; discount 6 per cent. per annum? Ans. $923.07.

EQUATION OF PAYMENTS.

THIS rule teaches us to find a mean time for the payment at once, of several debts due at different times, so that no loss of interest shall be sustained by either party.

RULE. Multiply each payment by its time, and divide the sum of the several products by the whole debt, and the quotient will be the equated time for the payment of the whole.

1. If I owe my neighbor $100 payable in 6 months; $120 payable in 7 months; and $160 to be paid in 10 months; when can the three sums be paid at once, without loss to either of us?

The interest on $100 for 6 months = int. on $600 1 mo.
"   "   120   7   "   = "   "   840 1 "
"   "   160   10 "   = "   "   1600 1 "

      ———   ———
      380   3040

Adding these products together, we find, that the interest on $3040 for 1 month is equivalent to the interest of the several sums for their respective times of payment. But $380 is the sum actually to be paid; and the question is, in what time it will produce the same amount of interest, that $3040 does in 1 month. If stated it would stand,

princ.   princ.   mo.
380   :   3040   ::   1   :   A.   But the third term being always unity, may be omitted, and the sum of the products be merely divided by the sum of the payments.
3040÷380=8.   Ans. 8 months.

2. If $400 are now due, $400 payable in 4 months, and $400 in 8 months; what is the equated time for paying the whole?   Ans. 4 months.

. 3. A man bought a farm, and agreed to pay $\frac{1}{4}$ of the price down, and the residue in three equal annual instalments; what is the equated time for paying the whole at once?
                          Ans. 18 months.

4. A owes B $600 to be paid in 2 years from the date of the note; but at the expiration of 6 months, A agrees to pay $150, if B will wait enough longer for the balance to compensate for the advance: how long ought B to wait?
                          Ans. 6 months

5. If $ 750 are to be paid $\frac{2}{3}$ of it in $1\frac{1}{2}$ years, $\frac{3}{10}$ of it in 2 years, and the residue in $2\frac{1}{2}$ years ; what is the equated time of paying the whole at once ?   Ans. $23\frac{2}{5}$ months.

6. If $\frac{1}{2}$ of a sum of money be now due, $\frac{1}{4}$ in 4 months, and the residue in 8 months ; what is the equated time of payment ?   Ans. 3 months.

PROFIT AND LOSS PER CENT.

WHEN commodities are bought and sold again, it is often desirable to know the profit or loss per cent. ; or, at what price they must be sold to gain a certain per cent.

1. If I buy factory cotton for 2s. per yard, and sell it at 2s. 8d. ; what do I gain per cent. ?

Here, the gain is 8d. on 2s., or 24d., and expressed fractionally, it is $\frac{8}{24} = \frac{1}{3}$. If this fraction be reduced to a decimal and carried to hundredths, (that is, to 2 places of decimals,) it will express the per cent.

$$3)100$$

Ans. $33\frac{1}{3}$ per cent.

2. If I buy broadcloth at $ 3.44, and sell it at $ 4.30 per yard ; what is the profit per cent. ?

Sold for - $ 4.30	.86	.43	172) 4300(25 per cent
Cost - 3.44	3.44 = 1.72	344
Gain - .86		860
		860

3. Bought cloth at $62\frac{1}{2}$ cts. per yard, and sold it at $ 1 ; what is the profit per cent. ?

$$\begin{array}{l} 1. \quad .375 \quad 3 \\ .625 \quad \overline{.625} = \overline{5} \end{array} \qquad 5)300$$

Profit - .375   Ans. 60 per cent.

RULE. Make the cost of the article the denominator, and the gain or loss the numerator of a fraction, and if there are decimals, make the number in both the terms equal : then reduce the fraction to its lowest terms, annex two ciphers to the numerator, and divide by the denominator.

4. If a merchant buy broadcloth at $ 5.50 a yard, and sell it at $ 6.60 a yard ; what is his profit per cent. ?

Ans. 20 per cent.

5. A man bought 500 sheep at $ 2.25 a head, and his expenses in the purchase were $ 75. He sold them again at an average price of $ 3.40 per head ; what was the profit per cent. on his investment? Ans. $41\frac{3}{4}$ per cent.

6. A grocer bought tea at 6s. a pound, but in consequence of a fall in the price of the article, is obliged to sell at 5s. 4d. per pound ; what is his loss per cent. ?

Ans. $11\frac{1}{9}$ per cent.

To know how a commodity must be sold, in order to gain or lose so much per cent.

RULE. Make a fraction of the per cent, and reduce it to the lowest terms : then take the parts of the purchase price indicated by the fraction, and add or subtract them according as it is gain or loss per cent.

7. If I buy Irish linen at 2s. 3d., how must I sell it per yard to gain 25 per cent. ?

$$25 \text{ per cent. is } \tfrac{25}{100} = \tfrac{1}{4}.$$

$$
\begin{array}{r}
\text{s.} \quad \text{d.} \\
\tfrac{1}{4})2 \quad 3 \\
\hline
6 \quad 3 \\
\hline
\end{array}
$$

Ans. 2s. 9d. 3 qr.

8. If I buy rum at $ 1.05 per gallon, how must it be sold to gain 30 per cent. ?

$$\tfrac{30}{100} = \tfrac{3}{10}$$

$$
\begin{array}{r}
\tfrac{3}{10})1.05 \\
.315 \\
\hline
\end{array}
$$

Ans. $ 1.365.

9. If tea cost 54 cents per pound, how must it be sold to lose $12\frac{1}{2}$ per cent. ?

$$\tfrac{125}{1000} = \tfrac{1}{8}$$

$$
\begin{array}{r}
\tfrac{1}{8}).54 \\
.06\tfrac{3}{4} \\
\hline
\end{array}
$$

Ans. $.47\frac{1}{4}$ cts.

10. If $ 126.50 are paid for 11 cwt. 1 qr. 25 lb. of sugar ; how must it be sold a pound to make 30 per cent. profit ?

Ans. $12\frac{1}{2}+$ cents.

11. If I buy 12½ cwt. of sugar for $140; at how much must I sell it per pound, in order to make 25 per cent. profit?
Ans. 12½ cents.

12. If a firkin of butter, containing 56 lb., cost $7; at how much must it be sold per pound to make 30 per cent profit?
Ans. 16¼ cents.

13. From a cask of wine containing 60 gallons, 5 gallons leaked out; at what price per gallon, must the residue be sold in order to gain 10 per cent. on the whole prime cost, it having been bought for $2.50 a gallon?
Ans. $3.

14. If by selling broadcloth at $3.25 per yard, I lost 20 per cent.; what was the prime cost?

$$\frac{20}{100} = \frac{1}{5}.$$

If I lost ⅕, then $3.25 must be ⅘ of the cost.

$$\frac{4}{5})3.25$$
$$81\frac{1}{4}$$

$4.06¼ Ans

15. If by selling muslin at 50 cts. a yard, I gained 25 per cent.; what was the first cost?

$$\frac{125}{100} = \frac{5}{4}$$

50 cts. = ⁵⁄₄ of the cost.
Ans. 40 cents.

16. If by selling cloth at $6.50 a yard, I lose 20 per cent.; what was the prime cost of it a yard?
Ans. $8.12½.

17. If by selling wheat at $1.36¼ a bushel, there be a profit of 30 per cent.; what was the original cost?
Ans. $1.05.

FELLOWSHIP.

By this rule, profits, losses, expenditures, assessments, &c., are apportioned among individuals, in the ratio of their respective contributions to a joint stock, their shares in the benefit of a joint privilege, or the relative amount of their property. It is not applicable to business partnerships, as generally conducted, in which a *fixed* proportion of the profit or loss is usully agreed on between the partners, and the disparity in the amount of capital compensated, in some cases, by allowance of interest on the excess furnished by either.

1. Three individuals associated for the purchase of a tract of land, for which they paid $ 10000. A furnished $ 4500, B $ 3500, and C $ 2000. They sold it at an advance of 50 per cent. on the purchase money; what was the share of each in the profits?

A's capital - 4500
B's " - 3500   50 per cent. on
C's " - 2000 $ 10000 = $ 5000.

10000 : 4500 : : 5000 : A's share = $ 2250.

B's share is found by making his capital the 2d term of the statement = $ 1750.

C's share in the same way, = $ 1000.

The last share may often be most conveniently found, by subtracting the other shares from the whole profit.

2. Three merchants make a joint stock of £ 1200, of which A put in £ 240, B £ 360, and C £ 600; and they gained £ 325.

The respective shares of the partners in the stock are $\frac{240}{1200} = \frac{1}{5}, \frac{360}{1200} = \frac{3}{10}, \frac{600}{1200} = \frac{1}{2}.$

Therefore A will have $\frac{1}{5}$ of 325l. = £ 65
B " " $\frac{3}{10}$ " " = £ 97 10s.
C " " $\frac{1}{2}$ " " = £ 162 10s.

Rule. Make the whole capital the denominator, and each man's share the numerator of a fraction; reduce it to its lowest terms, and take the parts of the whole profit or loss expressed by each of the fractions.

3. Several neighbors associated to maintain a select school, to consist of 30 scholars; each agreeing to pay in proportion to the number sent by him. They paid the teacher $ 35 per month, and the expenses amounted to $ 8.75 for 3 months. How much was it a scholar, and what had A to pay who sent 4 scholars?

Ans. $\begin{cases} \$ 3 .79 \text{ a scholar.} \\ \$ 15.16 \text{ for 4.} \end{cases}$

4. A man bequeathed to the eldest of his three children $ 3000, to the second $ 2500, and to the youngest $ 2000;

(12*)

but, on the settlement of estate, it was found to amount in all to but $ 6000. In what proportion ought it to be shared among the heirs, according to the tenor of the will ?

Ans. $\begin{cases} \text{\$ 2400.} \\ \text{\$ 2000.} \\ \text{\$ 1600.} \end{cases}$

5. A bankrupt failed for $ 20000, and his available means amounted in all to $ 13654 ; what will two of his creditors respectively receive, to one of whom he owes $ 3060, and to the other $ 1530 ?

Ans. $\begin{cases} \text{\$ 2089.06.} \\ \text{\$ 1044.53.} \end{cases}$

6. 500 barrels of flour were shipped for the West Indies ; 350 of them belonged to one individual and the residue to another. In stress of weather 100 of them were obliged to be thrown overboard ; how should the loss be shared between the two owners?

Ans. $\begin{cases} \text{The first} \quad \text{- 70 barrels.} \\ \text{The second 30 barrels.} \end{cases}$

7. Three men hired a pasture for $ 35, and one of them put in 4 cattle, another 7, and the third 9 ; what proportion of the expense, ought each to pay ?

Ans. $\begin{cases} \text{\$ 7.} \\ \text{\$ 12.60} \\ \text{\$ 16.40.} \end{cases}$

8. Four men hired a coach to convey them to their respective homes, which were at distances from the place of starting as follows : A's 16 miles, B's 24 miles, C's 28 miles, and D's 36 miles ; what ought each to pay, as his part of the coach hire, which was $13 ?

Ans. $\begin{cases} \text{A \$ 2.} \\ \text{B \$ 3.} \\ \text{C \$ 3.50.} \\ \text{D \$ 4.50.} \end{cases}$

9. If a tax of $ 3000 be assessed upon all the taxable property of a town, the valuation of which on the assessment roll, is $ 600000, what will it be on a dollar ; and what will A pay, who is assessed for $ 2500 property ?

$\frac{3000}{600000} = \frac{1}{200}$, which reduced to a decimal, $=.005$, or 5 mills on $ 1.

A's property - - 2500
.005
_____

Ans. $ 12.50.

In making up the tax list of a town, the amount of the polls (which pay a specific sum each,) is first to be deducted from the whole tax; then, the residue is assessed upon all the property, real and personal, of the town, as contained in an accurate inventory taken every year. Making the tax to be raised (after deducting the polls) the numerator, and the whole amount of taxable property the denominator of a fraction, we have the per cent. to be paid on a dollar.

It will facilitate the calculations, to form a table in the following manner. If, as in the last example, the tax amount to 5 mills on the dollar:

TABLE.

$ 1 pays .005	$ 10 pays .05	$ 100 pays .50
2 " .01	20 " .10	200 " 1.00
3 " .015	30 " .15	300 " 1.50
4 " .02	40 " .20	400 " 2.00
5 " .025	.50 " .25	500 " 2.50
6 " .03	60 " .30	600 " 3.00
7 " .035	70 " .35	700 " 3.50
8 " .04	80 " .40	800 " 4.00
9 " .045	90 " .45	900 " 4.50

$ 1000 pays $ 5.00.

The convenience of such a table will be seen from the application of it to an example.

10. What will B pay who is assessed for $ 3625 of taxable property ?

$ 3000 pays 15.
600 " 3.
20 " .10
5 " .025
———
Ans. $18.125.

11. What will C pay, whose property on the assessment roll is valued at $2593 ? Ans. $ 12.96½

12. The corporation of a village, for a public improvement, levies a tax of $ 540 on the property within its limits, the valuation of which is $ 270000.

Let the pupil find how much it is on a dollar, and construct an assessment table accordingly.

13. What will D pay, who is taxed for $ 2158 ?

Ans. $ 4.31½

14. Divide 240 into three parts, which shall be in pro
portion to each other, as the numbers 1, 2, and 3.

$$\text{Ans.} \begin{cases} 40 \text{ the 1st part} \\ 80 \text{ the 2d part} \\ 120 \text{ the 3d part.} \end{cases}$$

15. Three men hired a pasture, for the use of which
they agreed to pay 30 dollars. A put in 5 cows 12 weeks;
B 4 cows 10 weeks; C 3 cows 15 weeks: what share of
the rent ought each to pay?

5 cows 12 weeks=60 cows 1 week.
4 cows 10 " =40 " 1 week.
3 cows 15 " =45 " 1 week.

145

145 : 60 :: 30 : A's share - - - - - $12.41.
Making 40 the 3d term, we have B's share - $ 8.27½
In the same way we find C's share. - - - $ 9.31½

That is, when the consideration of time is involved in the
fellowship: *multiply each man's share by his time* ; *then, as
the sum of the products is to each particular product, so is the
amount to be shared to the share of each partner.*

16. Three men took a field of grain to harvest and thresh
on shares: A furnished four hands 5 days; B 6 hands 4
days; and C 8 hands 5 days. The whole crop amounted
to 630 bushels, of which they were to have one fifth.
What was the share of each partner?

$$\text{Ans.} \begin{cases} \text{A's 30 bushels.} \\ \text{B's 36 bushels.} \\ \text{C's 60 bushels.} \end{cases}$$

INVOLUTION.

A POWER is the product of a number, multiplied into
itself any number of times.

Involution is the raising of powers from any given

number as a root. Thus, if we involve the number 2, we have,

$2=$ the root or 1st power.

$2\times2=4^2$, the 2d power or square of 2.

$2\times2\times2=8^3$, the 3d power or cube of 2.

$2\times2\times2\times2=16^4$, the 4th power or biquadrate

$2\times2\times2\times2\times2=32^5$, the 5th power.

$2\times2\times2\times2\times2\times2=64^6$, the 6th power.

The figures on the right of the powers, are called indices, or exponents, and show how many times the root is taken as a factor in the production of the power.

The exponent is always 1 greater than the number of multiplications producing the power.

If we take the powers indicated by any two exponents, and multiply them into each other, their product will be the power indicated by the sum of the two exponents. Thus $4^2\times16^4=64$, the exponent of which is ^{6}: or, if we multiply any power into itself, the product will be the power indicated by the addition of the exponent to itself. Thus, $8^3\times8^3=64^6$. Likewise, if a higher power be divided by a lower, the quotient will be the power indicated by the difference of the exponents. Thus, $64^6\div16^4=4^2$. $(^6-^4=^2.)$

1. Involve 8 to the 4th power.   Ans. 4096.
2. What is the 2d power of 45 ?   Ans. 2025.
3. What is the 3d power of 3.5 ?   Ans. 42.875.
4. What is the 4th power of 1.2 ?   Ans. 2.0736.
5. Find the square of $\frac{2}{3}$.   Ans. $\frac{4}{9}$.
6. Find the 3d power of $\frac{5}{9}$.   Ans. $\frac{125}{729}$.

EVOLUTION, OR THE EXTRACTION OF ROOTS.

EVOLUTION is the reverse of Involution. It is the operation by which, having a power given us, we find the root which, by multiplication, produced the power.

Although there is no number but will produce a perfect power by involution, yet there are many numbers of which precise roots can never be found. But, by the help of decimals, we can approximate towards the root to any degree of exactness.

The roots which approximate are called surd roots, and those which are exact are called rational roots.

A Table of the Squares and Cubes of the Nine Digits.

Roots.	1	2	3	4	5	6	7	8	9
Squares.	1	4	9	16	25	36	49	64	81
Cubes	1	8	27	64	125	216	343	512	729

The following radical sign is used to express the extraction of roots : $\sqrt{}$. Thus, $\sqrt{9}=3$, denotes that the square root of 9 is 3. To indicate the roots of powers higher than the square, their exponent is placed above the radical sign. Thus, $\sqrt[3]{}$ expresses the cube root, and $\sqrt[4]{}$ the biquadrate. Roots are also indicated by a fraction, with the exponent of the power for a denominator. Thus, $(64)^{\frac{1}{2}}$, $(64)^{\frac{1}{3}}$ express the square and cube root of 64.

EXTRACTION OF THE SQUARE ROOT

THE name Square Root, is derived from a particular application of the rule, in finding the side of a square from its area or superficial contents.

Any number multiplied into itself, produces a square, and the object of the square root is, to find the number, which, being thus multiplied, will produce the given square.

To explain the method by which this is done, let us take a number and multiply it, and by tracing back the operation, endeavor to educe the principle of the rule.

We will take the number 36, and (as it makes no difference in the result, in what order we multiply,) will begin with the tens, and multiply 3 by the 3, (30 by 30.)

$$
\begin{array}{r}
36 \\
36 \\
\hline
\end{array}
$$

30 multiplied by 30 - - - - - - 900
6 multiplied by 30 - - - - - - 180
And again, 30 in the upper number, multiplied by 6 - 180
Lastly, 6 multiplied by 6 - - - - - - - 36

Sum of the products, or square of 36   - -   1296

Now, if we did not know the root which produced this power, how should we proceed to find it?

We begin with finding the *highest* figure in the root first: in order to which, it is evident we must take that portion of the power, where the product of the highest figure is found; and so likewise, in finding the inferior figures of the root, we must at each successive step, take those places in the power, where their respective products are situated.

Now, if we take any number and involve it to the square, as a unit, a ten, a hundred, and so on, we shall discover, that at each remove of the factor towards the left, its *proper product* (that is the product of the figure by itself) falls two places towards the left. Thus,

$$1 \times 1 = 1 \\ 10 \times 10 = 100 \\ 100 \times 100 = 10000$$ and $$9 \times 9 = 81 \\ 90 \times 90 = 8100 \\ 900 \times 900 = 810000$$

Indeed, we have but to reflect, that at each remove, two additional ciphers are thrown into the product. We see, therefore, that the product of a unit by itself, falls in the units place, and can at most influence but *one* place above; the product of the tens falls in the place of hundreds, and may extend to one place above. So also of the hundreds. As a preliminary, therefore, to the evolution, we designate the places in the power, *where the first figure in the product of each of the factors of the root falls*, by putting a point over the unit's place, and over every alternate figure towards the left. This divides the given number into periods, each of which contains the proper product of one of the figures in the root, besides what was carried for the multiplication of it, into the next figure below it.*

Now, the first period at the left of the power, must be the product of the first figure in the root into itself, increased by what was carried from the other figure. We, therefore, seek what figure multiplied into itself will produce that

---

* The reason generally assigned for placing the point, both in the square and cube roots, is, that it shows *the number of figures of which the root will consist.* But it will as well do this, if we begin at the *highest* as at the lowest place of the given number. The true reason is, that in the evolution of the root, we must take in succession those portions of the power, which contain the products of the corresponding figures of the root.

period. We see that 3 comes nearest to it, (for $4 \times 4 = 16$, which is too much,) and we multiply it into itself, (for it was so multiplied in producing the power,) and have $3 \times 3 = 9$, or, (because it occupies the ten's place,) $30 \times 30 = 900$.

$$1296(3$$
$$900$$

Subtracting this power, we have remaining $\overline{396}$

We next wish to find what factor produced (by multiplication) this remainder.

We find (by looking at the preceding analysis) that it was multiplied twice into the figure which we have found, (3,) and once into itself. Now, multiplying any number twice by the same factor, is equivalent to multiplying *twice* the number *once*. If, therefore, we double the root already found, and divide the remainder by it, the quotient will be the figure, which produced the number 396; (allowing for what is to be carried for the multiplication of the number into itself.)

Twice 3 (30)$= 6(60)$ and 60 into 396, 6 times. But this quotient figure was multiplied into itself, and we, therefore, place it at the right of the divisor in the unit's place, making 66.

$$66)396(6$$
$$396$$

We find the whole root to be **36**, the number which produced the power.

RULE.—1. Distinguish the given number into periods of two figures each, by putting a point over the place of units, another over the place of hundreds, and so on; and if there are decimals, point them in the same manner, from units towards the right hand.

2. Find the greatest square in the first, or left hand period, place the root of it at the right hand of the given number for the first figure of the root, and the square itself under the period, and subtract it therefrom, and to the remainder bring down the next period for a dividend.

3. Double the root already found, and annex a cipher to the product, for a divisior. Seek how many times this di-

visor is contained in the dividend, place the result in the quotient, and also in the cipher's place in the divisor, and multiply by it. Subtract the product from the dividend, bring down the next period and form a new divisor in the same manner as before.

If any dividend be found too small to contain the divisor, a cipher must be placed in the quotient.

If there be a remainder at the end of the operation, it may be carried into decimals, by adding two ciphers at each division.

1. Required the square root of 141225.64.

$$141\overset{\cdot}{2}25.\overset{\cdot}{6}4(375.8\dagger \text{ Ans.}$$
$$9$$

Double the root (3)=6	67)512
and with a cipher annexed=60*	469

Double the root (37)=74	745)	4325
and with a cipher annexed=740		3725

7508)60064
60064

2. Find the square root of 2025. Ans. 45.
3. Find the square root of 17.3056. Ans. 4.16
4. Find the square root of .000 729. Ans. .0 27
5. Find the square root of 566.44 Ans. 23.8
6. Find the square root of 45. Ans. 6.708
7. Find the square root of .002916. Ans. .054

To extract the square root of vulgar fractions.

RULE. Reduce the fraction to its lowest terms. Extract the root of the numerator for a new numerator, and the root of the denominator, for a new denominator.

---

* As the cipher is to be replaced by the next quotient figure, it need not actually be set down but merely considered as forming a part of the divisor.

† If this root be multiplied into itself, it will produce the power.

If the fraction be a surd, reduce it to a decimal, and extract its root.

Mixed numbers must be reduced to improper fractions before their root is extracted; or, the fractional part of them may be reduced to an equivalent decimal, which being annexed to the integer, the root of the whole may be extracted.

8. What is the square root of $\frac{25}{36}$?   Ans. $\frac{5}{6}$

9. What is the square root of $\frac{4}{7}\frac{7}{7}$,   Ans. $\frac{4}{7}$.

10. What is the square root of $\frac{9}{12}$?   Ans. .866025

11. Find the root of $17\frac{3}{8}$.   Ans. 4.168333

12. Find the root of $20\frac{1}{4}$.   Ans. $4\frac{1}{2}$.

PROPERTIES OF A SQUARE.

DEFINITION. I.   An angle is formed by the meeting of two straight lines.

II. If the meeting lines are perpendicular to each other, the angle formed is a right angle.

III. A square is a figure formed by four equal sides, and containing four right angles

IV. A straight line connecting two opposite angles of a square, is called a diagonal.

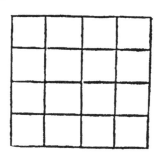   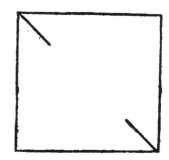

V. A rectangle or oblong, is a figure, whose opposite sides are parallel, and its angles all right angles.

PROPOSITION. The area or superficial content of any rectangular figure is found by multiplying the length into the breadth.

If for example, the divisions of the sides of the foregoing square, represent inches, there are 4 of them in length, and 4 in breadth, and $4 \times 4 = 16$: on counting the minor squares there will be found to be 16 of them.

The area of a square then, is equal to the product of one of its sides into itself; in other words, to the *square of the side.*

Thus, if one side of a square field be 40 rods, the area of the field $= 40 \times 40 = 1600$ square rods.

Now, as we find the area, or superficial content of a square, by squaring one of its sides, we find the side by the converse of this operation ; that is, by extracting the square root of the area.

Suppose, for example, you have 2025 apple-trees, to set out in a square field, and you desire to know how many you must plant in a row, to have an orchard in a square form. The square root of 2025 is 45, the number of trees in a row.

Suppose you have 512 trees to be set out, and wish the rows to be twice the length in one direction, that they are in the other. Take half the number of trees (256), and extract the square root of it, which will dispose one half of your trees in a square, and give you the length of the shortest side. Double that number for the longest side.

Ans. 16 trees one way ; 32 the other.

13. A section of land in the Western states is a square, consisting of 640 acres ; what is the length in rods of one of its sides ? Ans. 320 rods.

14. A certain square pavement, contains 20736 square stones ; I demand how many stones there are in one of its sides. Ans 144.

15. What must be the side of a square field, that shall contain an area equal to another field of rectangular shape, the two adjacent sides of which are 18 rods and 72 rods ? Ans 36 rods.

16. Suppose 3097600 men to be drawn up in a solid square, how many men would there be on a side ? and, allow-

ing each man to occupy a square yard of ground, how large a plain would contain the whole number ?

$\begin{cases} \text{Ans. 1760 men on a side.} \\ \text{1 mile square.} \end{cases}$

DEFINITION I. A triangle is a figure of three sides, and having three angles.

II. If one of its angles be a right angle, the figure is called a right-angled triangle.

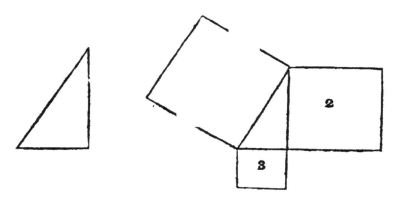

By a geometrical demonstration it is proved, that the square formed on the longest side of a right-angled triangle (that opposite the right angle,) is equal to the sum of the squares formed on the other two sides : that is, its area is equal to the areas of the other two squares. Thus the square 1, is equal to the squares 2 and 3 in the diagram above.

On this curious relation of the sides of a right-angled triangle a great many very interesting and useful calculations are based.

If for example, we know the length of the two shorter sides of such a figure, we can calculate that of the longest side (or hypotenuse,) without measuring it. We have only to square each of the two sides, add their squares together, and extract the square root of the sum.

17, Suppose for example, a surveyor has run a line due north 60 rods, and then east 80 rods, and wishes to know his distance in a right line from his first station.

$$60 \times 60 = 3600$$
$$80 \times 80 = 6400$$

Sum of the two squares   -   -   -   -   10000

Square root of this sum   -   -   -   -   100 rods Ans.

18. If the room where you are sitting, be 28 feet long and 22 wide, what is the distance between its opposite angles.

Ans. 35.6+feet.

19. If from the ground to the eaves of a house, the perpendicular height be 30 feet, and you wish to make a ladder, which being planted at the distance of 10 feet from the wall will reach the roof, of what length must it be?

Ans. 31 feet 7 in.

20. A carpenter wishes to connect a post and beam by a brace, which shall enter a mortise in each, at the distance of 4 feet on the inside from the junction or angle. Of what length must the brace be to the entrance of the upper side of the mortise?

Ans. 5 feet 7.8 inches.

21. A hawk perched on the top of a perpendicular tree, 77 feet high, was brought down by a sportsman standing at the distance of 14 rods, on a level with its base; what distance in yards did he shoot?

Ans 81.15+yards.

As the square of the hypotenuse equals the sum of the squares of the two sides, it is evident, that the square of one of the two sides, is equal to the square of the hypotenuse less the square of the other side.

22. If a ladder 40 feet long be so planted as to reach a window 33 feet from the ground, on one side of a street, and by turning it over, without moving its foot it will reach a window 21 feet high on the other side, how wide is the street?

Here the ladder represents the hypotenuse, and the height of the window one of the sides of a right-angled triangle. The distance of the foot of the ladder from the wall, the required side.

$40 \times 40 = 1600$ - - square of the hypotenuse.
$33 \times 33 = 1089$ - - square of the known side.

511 - - - square of the required side.

The square root of 511 will be the distance from the foot of the ladder to the wall on one side; which, added to the distance to the wall on the other side, found in the same way, will give the entire width of the street.

Ans. 56.649 feet.

23. Required the height of a May-pole, whose top be-

(13*)

ing broken off by a blast of wind, struck the ground at the distance of 15 feet from its foot, and measured 39 feet.

Ans. 75 feet

24. Wishing to know the distance from a station near the side of a river to a tree below, but prevented by a curve in the stream from measuring it, I run out at right angles to the line of direction 15 rods, and then 30 rods in a right line to the tree. What was the distance required?

Ans. 26 rods, nearly.

25. If the distance from a point perpendicularly under a kite flying in the air, to the station of the boy holding the string, be 300 feet, and the length of string from the hand to the kite be found to measure 580 feet; how high is the kite above the surface of the ground?

Ans. 496 feet 4 inches.

NOTE.—In this case the string must be held to the ground when the measurement is made.

26. If the diagonal of a rectangular field measure 40 rods, and one of its sides 32 rods; what is the length of the adjacent side?   Ans. 24 rods.

27. If the diagonal of a square field be 60 rods; what is the length of a side, and what the area of the field?

$\begin{cases} \text{Length of a side 42.42 rods.} \\ \text{Area of the field 1800 square rods.} \end{cases}$

NOTE. — The diagonal being the hypotenuse of a right-angled triangle, of which the other sides are equal, its square is double the square of each of the sides.

28. Suppose the width of a barn to be 36 feet, of what length must the rafters be, if the elevation of the ridge above the level of their foot be 14 feet?   Ans. 22 feet 9½ in.

NOTE. — Each rafter is the hypotenuse of a triangle, and the height of the ridge one of its sides.

29. If 4200 fruit trees are to be planted, how large a square can be formed out of the number; that is, how many trees will there be in a row?   Ans. 64 trees.

The square of 64=4096; consequently there will be 104 trees *over*, which cannot be united to the square, without destroying its figure.

There must always be a remainder, unless the given number be an *exact power*.

To find the side of a square that shall bear any assignable ratio to a given square :

RULE.—Multiply the given square by the ratio of the required square, and extract the square root of the product.

30. If the side of a square field be 40 rods ; what will be the side of another square field that shall contain just twice the quantity ?

$40 \times 40 = 1600 \times 2 = 3200$. $\sqrt{3200} = 56\frac{1}{2} +$ rods Ans.

31. What would be the side of a field that would contain one half the quantity ?   Ans. 28.4 rods

NOTE.—Multiplying by $\frac{1}{2}$ is simply taking one half the multiplicand.

To find the diameter of a circle that shall have any assignable ratio to a given circle.

RULE.—Multiply the square of the diameter, by the ratio of the required circle, and the square root of the product will be the diameter of the required circle.

32. If the bottom of a circular cistern be 5 feet in diameter ; what is the diameter of another cistern, whose bottom contains just 4 times as much surface ?

Ans. 10 feet.

QUESTIONS.

What is involution ?
What does the exponent of a power indicate ?
What relation has evolution to involution ?
What is the extraction of the square root ?
What is meant by the *proper product* of a figure ?
What is the use of the points placed over the number whose root is to be extracted ?
Over which places in the given number is it put ?
How may you prove the correctness of your work ?
Can an exact root be found to every number ?
What is a right-angled triangle ?
What relation is there between the square of the hypotenuse, and the squares of the two sides ?
If the two sides are given, how do we find the hypotenuse ?
If the hypotenuse and base are given, how is the perpendicular found ?
If the hypotenuse and perpendicular are known, how do we find the base ?

EXTRACTION OF THE CUBE ROOT.

Under the head of involution, it has been shown, that a cube results from multiplying a number into itself, and that product again by the same multiplier.

The extraction of the cube root is the operation by which we find the root, that is, a number, which, by involution will produce the given number or power.

In performing this, as in extracting the square root, we reverse the process of involution, and begin with finding the last figure employed in the multiplication. In order to this, it is necessary to designate the place in the series, where the *proper product* of each figure terminates.

Now, if we take any figure, (say 9,) and involve it successively, as units, tens, hundreds, and so on, we shall readily discover the law which governs the position of the successive products.

$9 \times 9 \times 9 = 729$. As a unit, its product terminates with the unit's place, and influences but two other places, the tens and hundreds. As a ten, its proper product is the same, but is removed three places toward the left: thus, $90 \times 90 \times 90 = 729000$. As a hundred, its product will fall three places higher :

$$900 \times 900 \times 900 = 729000000.$$

Now, if we take the lowest of the nine digits, (1,) and involve it, its proper product will terminate in the same places ; that is, three places higher for each remove toward the left.

$$1 \times 1 \times 1 = 1. \quad 10 \times 10 \times 10 = 1000.$$

$$100 \times 100 \times 100 = 1000000.$$

The termination, therefore, of the proper product of each figure of the root, is designated in the power, by a point over the unit's place, the thousand's place, and so on, assigning 3 places to each figure.

Before explaining the process of evolution, we will involve a number analytically.

$$24 = 20 + 4.$$

```
   24                          20+ 4
   24                          20+ 4
 ----                         ------
   96                          80+16 ⎫
   48                        400+80  ⎬ = the square.
 ----                                ⎭
  576
   24                          20+ 4

 2304                         320+64 ⎤
 1152                    1600+320    ⎥
 ----                    1600  320   ⎬ = the cube.
13824           8000+1600            ⎦
```

The constituent parts of this process are,

1. The cube of the 1st term - - - - 8000＝8000
2. Three times its square multiplied by the
    2d term - - - - - - - - - 1600＝4800
3. Three times the 1st term multiplied by
    the square of the 2d - - - - - 320＝ 960
4. The cube of the 2d term - - - - 64＝ 64

$$(24)^3 = 13824$$

In extracting the root of this number, we commence with designating by a point, the places where the *proper product* of each figure terminates.

The first period contains the cube of the 1st figure, increased merely by what was carried for the previous multiplication. The greatest cube in 13 is 8, the root

$$2^3 = 8$$

$$\overset{\cdot\quad\cdot}{13824}(24$$

$$2^2 \times 3 = 12)\,58 \text{ - dividend.}$$

First two periods - - $\overset{\cdot\quad\cdot}{13824}$

$$(24)^3 = 24 \times 24 \times 24 = 13824$$

of which (2) is to be placed in the quotient. The cube itself being subtracted, we bring down the first figure in the next period for a dividend, and for a divisor multiply the square of the quotient by 3.

The reason of dividing by 3 times the square of the root, is found in the fact, that the remainder of any period, after subtraction, and the first figure in the next period, contains the product of 3 times the square of the previous figures of the root into the next figure of the root.*

RULE 1.—Point off the given number into periods of 3 figures each, beginning at the units.

2. Find the greatest cube in the left-hand period, and place its root in the quotient. Subtract the cube from the first period, and to the remainder bring down the first figure of the next period for a dividend.

3. Multiply the square of the quotient by 3 for a divisor ; divide, and place the result in the, quotient. Cube the quotient, and subtract the result from the first two periods, and to the remainder bring down the first figure of the 3d period for a new dividend.

4. Proceed as before to find a new divisor, &c., and in all cases subtract the cube of the whole quotient, from as many of the left-hand periods, as there are figures in the root already found.

If there be a remainder, at the last subtraction, the operation may be carried into decimals, by adding periods of ciphers at pleasure.

If there be decimals in the given number, the periods must be rendered complete, by the addition, if necessary, of ciphers.

When the given number consists wholly of decimals, the periods are pointed from left to right; the first point falling on the place of thousandths, and so on.

If the given number be a vulgar fraction, or mixed number, after having reduced it to its lowest terms, the root of the numerator and denominator may be extracted separately; or, if the terms do not contain an exact root, the fraction may be reduced to an equivalent decimal, and the root be then extracted.

---

* By looking at the analysis, this product will be seen to be 48|00, (the ciphers falling into the two inferior places of the power are not included in the dividend,) which becomes 58, by the addition of the highest place in the product of the square of the next figure, by three times the last figure in the root.

As the dividend contains *more* than the product of the divisor and the next quotient figure, care must be taken not to make that figure *too large*.

1. What is the cube root of 32461759?

$$32461759(319$$
$$3^3 = 27$$

$3^2 \times 3 = 27^*) \ 54$ dividend.

First two periods - - 32461
$(31)^3 = 31 \times 31 \times 31$ - - - 29791

$(31)^2 \times 3$ - - - - - 2883) 26707 - - 2d dividend.

First three periods - - 32461759
$(319)^3$ - - - - - - - 32461759

Ans. 319.

NOTE.—If, in any case, a cube be found to be greater than the periods from which it is to be subtracted the last quotient figure must be made less.

2. Find the cube root of 48228.544   Ans. 36.4.
3. Find the cube root of 205379.   Ans. 59.
4. Find the cube root of 162.771336.   Ans. 5.46.
5. Find the cube root of 21024576.   Ans. 276.
6. Find the cube root of $\frac{8}{64}$.   Ans. $\frac{2}{4}$.
7. Find the cube root of $\frac{343}{729}$.   Ans. $\frac{7}{9}$.
8. Find the cube root of .000684134.   Ans. .088+.
9. Find the cube root of 2.   Ans. 1.25+.
10. Find the cube root of $\frac{4}{5}$.   Ans. .92+.
    Change the fraction to a decimal.
11. Find the cube root of 65890311.319. Ans. 403.9.

APPLICATION OF THE CUBE ROOT.

DEFINITION.—A geometrical cube is a solid contained by six equal sides, all of which are squares, and its angles right angles.

The contents of such a solid are found by multiplying

---

* The divisor is contained exactly twice, but allowance must be made for the influence of other figures in the production of the dividend.

together its three dimensions of length, breadth, and thickness ; or, (which is the same thing) cubing one of its sides.

If, for example, a cubic foot be made up of detached
cubic inches ; in one side there will be 12 rows of these
inch blocks, with 12 in a row : $12 \times 12 = 144$, the number in one side. There are also 12 in thickness, and
$144 \times 12 = 1728$, the number of solid inches in the whole
block.

On the other hand, if the contents of a cubical body
are given, its side is found by extracting the cube root.

12. What is the side of a square block of marble, that
shall contain 13824 cubic inches ? Ans. 24 inches.

13. The pedestal of a monument was a square block of
granite, containing 373248 cubic inches ; what was the
length of one of its sides ? Ans. 6 feet.

14. The statute bushel contains 2150.4 cubic inches ;
what is the side of a cubic box that will contain 50 bushels ?
Ans. 47.5 inches.

PROPOSITION. The solid contents of cubes or spheres are
to each other, as the cubes of their like dimensions.

15. If a block of marble, 2 feet square weigh a ton, how
much would it weigh if its sides were 8 feet square ?

$$2^3 = 8 \qquad 8^3 = 512.$$

$$8 : 512 :: 1 : A. \qquad \text{Ans. 64 tons.}$$

The above is the method of solution commonly given :
—a better one is as follows ; $8 \div 2 = 4$; $4^3 = 64$. That is,
*divide the dimension of the required solid by that of the given
one, and cube the quotient.*

16. Admitting the diameter of the earth to be 8000
miles, what would be its comparative magnitude, if it had
been 10000 miles in diameter ? Ans. $1\frac{61}{64}$

17. If a globe of silver of 3 inches diameter, be worth
$150, what would be the value of one 6 inches in diameter?
Ans. $1200

18 What must be the side of a cubical bin to hold $100\frac{1}{2}$.
bushels ? Ans. $60\frac{1}{10}$ inches.

19. What must be the side of a square cistern, to contain
600 gallons, allowing 282 cubic inches to the gallon ?
Ans. $55\frac{3}{10}$ inches.

20. If you have a pile of wood 32 feet long, 4 feet wide,

and 4 feet high ; what would be the side of a square pile containing the same quantity ? Ans. 8 feet.

21. If you have a cubical box 2 feet square, what must be the dimension of another cubical box to contain just 8 times as much ? Ans. 4 feet.

NOTE.—Multiply the cubical contents by the proportion of the required cube, and extract the root of the quotient.

22. If your cistern is 4 feet square, and you would enlarge it so as to contain 4 times as much water ; what must you make each of its sides ? Ans. 6 feet 4+in.

There is very little occasion for extracting the roots of powers higher than the cube. The 4th root may be found by extracting the square root twice : Thus, the 4th root of $16=2$; that is $\sqrt{16}=4$, and $\sqrt{4}=2$. The 5th root $=$ the square root of the cube root, or, the cube root of the square root. Thus $(64)^{\frac{1}{6}}=2$: that is, $(64)^{\frac{1}{3}}=4$; and $\sqrt{4}=2$: or $\sqrt{64}=8$; and $(8)^{\frac{1}{3}}=2$.

FOR EXTRACTING THE ROOTS OF ALL POWERS.

RULE 1. Point off the given number into periods consisting of figures equal in number to that indicated by the exponent* of the given power.

2. Find the first figure of the root by trial, and subtract its power from the left hand period of the given number.

3. To the remainder bring down the first figure in the next period, and call it the dividend.

4. Involve the root to the next inferior power to that which is given, and multiply it by the number denoting the given power, for a divisor.

5. Find how many times the divisor may be had in the dividend, and the quotient will be another figure of the root.

6. Involve the whole root to the given power, and subtract it (always) from as many periods of the given number as you have found figures in the root.

---

* For the square root 2 figures ; for the cube root 3 ; for the 4th root 4 figures, and so on.

(14)

7. Bring down the first figure of the next period to the remainder for a new dividend, to which find a new divisor as before, and in like manner proceed till the whole be finished

ALLIGATION.

ALLIGATION is compounding or mixing together several simples of different rates or qualities, so that the composition may be of some intermediate rate or quality.

CASE I. When the quantities and rates, or qualities, of several things are given, to the find rate or quality of the mixture.

RULE. Find by multiplication the value of each of the ingredients, and divide the sum of the values by the sum of the ingredients.

1. A farmer mixed 15 bushels of rye at 64 cents a bushel; 18 bushels of corn at 55 cents a bushel, and 21 bushels of oats at 28 cents a bushel: What is a bushel of the mixture worth?

bu		cts.		
15	at	64	- -	9.60
18	"	55	- -	9.90
21	"	28	- -	5.88

54          $25.38

54)25.38(.47 cts. Ans.

2. If 120 bushels of wheat be bought at 80 cents a bushel, and 75 bushels at 86 cents a bushel, what is the average cost per bushel of the whole purchase?

Ans. 82+cents.

3. An innkeeper mixed 13 gallons of water with 52 gallons of brandy, which cost him $1.25 per gallon; what is the value of 1 gallon of the mixture, and what his profit on the sale of the whole at 6¼ cents per gill?

Ans. $\begin{cases} \$1. \text{ a gallon.} \\ \$65. \text{ profit.} \end{cases}$

4. A has 50 shares of stock in four different banks ; viz. 15 shares in a bank, the stock of which is at an advance of 6 per cent ; 15 of stock which is at 4 per cent. advance ; 12 of stock which is $3\frac{1}{2}$ above par, and 8 shares of stock which is $5\frac{3}{8}$ per cent. above par. What is the mean per cent. advance on the whole, the nominal value of the shares being the same ?   Ans. $4\frac{7}{10}$ per cent.

5. At a given hour of the day, the thermometer for 2 days in the week stood at 70 degrees ; for 2 other days at 74 degrees ; 1 day at 77, and 2 days at 80 degrees ; what was the mean temperature for the week ?   Ans. 75 degrees.

CASE II. To find what quantity of any number of simples whose rates are given, will compose a mixture of a given rate or quality.

6. If a grocer have sugars worth 11cts. 13cts. 14cts. 15cts. and 16cts. per pound ; in what proportions must he mix them, in order that the mixture may be worth 12 cents per pound ?

It is evident that 1 pound at 11, may be put with 1 pound at 13 cts. because the excess of the one exactly balances the deficiency of the other : but 1 lb. at 14 cts. will require 2 at 11 cts. to balance it ; 1 lb. at 15 cts. will require 3 at 11 ; and 1 at 16 cts. will require 4 at 11 cts.

Had the price of the lowest sort been 10 instead of 11 cts., to preserve the balance, the quantities of those above the mean rate, would have to be proportionally increased, and instead of 1 pound of each being taken, there must be 2 pounds. Hence it is evident, that the greater the difference between the value of any one of the simples and the mean value of the mixture, the greater the quantity to be taken of the sort by which it is to be balanced, and that the difference between the rate of each and the mean rate, is the number to be taken of the other.

Thus, to form a mixture according to the conditions of the foregoing example :

cts.			cts.	cts.			cts.
1 lb. at 11	balances	1 at	13	1 lb. at 10	balances	2 at	13
2 lb. " 11	"	1 "	14	2 lb. " 10	"	2 "	14
3 lb. " 11	"	1 "	15	3 lb. " 10	"	2 "	15
4 lb. " 11	"	1 "	16	4 lb. " 10	"	2 "	16

Therefore, in the first case, there must be 10 lb. of the lowest rate, to 1 lb. of each of the other rates ; and in the second case, there must be 10 lb. of the lowest rate, to 2 lb. of each of the other rates.

RULE.—Set the rate of the several ingredients under each other, and connect each one less than the mean rate, with any one or more than one, greater than the mean rate ; and each one greater with one, or more than one of a less rate.

Place the difference of each and the mean rate against the ingredient with which it is connected.

If only one difference stand against any rate, it will be the required quantity of that ingredient; but if there be more than one, their sum will be the quantity required of that ingredient.

7. What proportions of gold 18 carats fine, 20 carats, 23, and 24 carats fine* must be taken to compose a mixture 22 carats fine?

$$22 \begin{cases} 18 \\ 20 \\ 23 \\ 24 \end{cases} \quad \begin{array}{llll} 2 \text{ grains} & 18 \text{ carats fine} \\ 1 \quad `` & 20 \quad `` \quad `` \\ 2 \quad `` & 23 \quad `` \quad `` \\ 4 \quad `` & 24 \quad `` \quad `` \end{array}$$

The ingredients might be alligated differently; as, the 18 with 23, and the 20 with 24 ; or all above the mean rate with *one* below. The result would be different in every instance, and yet be correct.

The rule is of so little practical use, that it is needless to multiply examples.

CASE. III. When one of the ingredients is limited to a certain quantity :

RULE.—Find the quantity of the several ingredients, as in the last case. Then divide the whole of the quantity limited, by the difference standing against its rate, and multiply the quotient by each of the other differences, for the quantity of the several rates.

8. A grocer has tea at 50 cents per lb. ; some at 85 cents, some at 90 cents, which he wishes to mix with 90 lb. worth

---

* Pure gold is 24 carats fine.

† Penny weights or any other denomination.

40 cents per lb., so that the mixture shall be worth 60 cents per lb. How much of the first three sorts must be taken?

$$60\begin{cases}40——\\50——\\85——\\90——\end{cases}\quad\begin{matrix}30\\25\\10\\20\end{matrix}\quad\begin{matrix}90\div30=3\\25\times3\ =75\\10\times3\ =30\\20\times3\ =60\end{matrix}\quad\begin{matrix}75\text{ lb. at }50\text{ cts}\\30\ ``\ ``\ 85\text{ cts.}\\60\ ``\ ``\ 90\text{ cts.}\end{matrix}$$

9. What quantity of rye at 48 cents, of corn at 36 cents, and of barley at 30 cents a bushel, being mixed with 10 bushels of wheat worth 70 cents a bushel, will form a mixture worth 38 cents a bushel?

Ans. $\begin{cases}2\frac{1}{2}\text{ bushels of rye.}\\12\frac{1}{2}\quad``\quad``\text{ corn.}\\40\quad``\quad``\text{ barley.}\end{cases}$

CASE IV. When the whole composition is limited to a given quantity:

RULE.—Find the quantity of the several ingredients as in Case II. Then, by the sum of the several quantities thus found, divide the given quantity, and multiply the quotient by each of the quantities severally.

10. A goldsmith has gold of 15, 17, 20, and 22 carats fine, and would melt together of all these sorts, so much as to make a mass of 40 oz. 18 carats fine. How much of each sort is required?

$$18\begin{cases}15——\\17——\\20——\\22——\end{cases}\quad\begin{matrix}4\text{ oz.}\\2\ ``\\1\ ``\\3\ ``\end{matrix}\ \Big\}\ \begin{matrix}\text{Whole quantity}\\40\div10=4.\end{matrix}\ \begin{cases}\begin{matrix}\text{oz.}\qquad\text{carats.}\\4\times4=16\text{ of }15\\4\times2\ ``\ \ 8\ ``\ 17\\4\times1\ ``\ \ 4\ ``\ 20\\4\times3\ ``\ 12\ ``\ 22\end{matrix}\end{cases}$$

Sum - 10 oz.

Different results may be obtained by alligating the simples differently.

11. Suppose a broker have stocks, some at 3 per cent. advance; some at 6 per cent.; some at 8 per cent.; and some at 10 per cent. advance; how many shares of each must he take, to make an aggregate of 54 shares, worth 7 per cent. advance?

Ans. $\begin{cases}18\text{ shares at }-\ 2\text{ per cent.}\\6\quad``\quad``\ -\ 6\text{ per cent.}\\5\quad``\quad``\ -\ 8\text{ per cent.}\\24\quad``\quad``\ -\ 10\text{ per cent.}\end{cases}$

(14*)

ARITHMETICAL PROGRESSION.

A SERIES of numbers increasing or decreasing by a common difference, constitutes an arithmetical progression.

When the numbers increase by a uniform addition, it is called an *ascending series;* when they decrease by a uniform subtraction, a *descending series.*

Thus, 2, 4, 6, 8, 10, is an ascending series, by the uniform addition of 2; and 10, 8, 6, 4, 2, a descending series, by the uniform subtraction of the same number.

The number added or subtracted is called the *common difference.*

The numbers themselves are called *terms,* and the first and last terms *extremes.*

PROPOSITION I. The sum of the extremes is equal to the sum of any two terms equidistant from them.

Thus, in the series 2, 4, 6, 8, 10; 2+10=12, and 4+8=12. The reason is, that from the constitution of the series, one of the mean terms is just as much greater than one of the extremes, as the other term is less than the other extreme.

PROPOSITION II. The difference of the extremes is equal to the common difference multiplied by the number of terms less 1.

Thus, in the series 1, 4, 7, 10, 13, 16; the common difference is 3 and the number of terms 6: $5 \times 3 = 15 = 16 - 1$.

As the common difference is added or subtracted to form every term *after* the 1st, the reason of this proposition is evident.

PROPOSITION III. The sum of all the terms of the series, is equal to the sum of the extremes multiplied by half the number of terms; or, multiplied by the number of terms, and the product divided by 2.

For example, 1, 3, 5, 7, 9, 11, 13, 15; $\overline{15+1} \times 4 = 64$.

The truth of the proposition will be made evident, if we

place the series inverted under itself, and add the correspondent terms.

Given series	-	1,	3,	5,	7,	9,	11,	13,	15
Inverted ditto	-	15,	13,	11,	9,	7,	5,	3,	1

$$16, 16, 16, 16, 16, 16, 16, 16$$

Here, we have as many times the sum of the extremes, as there are terms; and it is evident, that the series of equal terms is double the given series.

PROBLEM I.—The first term of a series, the last term, and the number of terms being given, to find the sum.

RULE.—Multiply the sum of the extremes by half the number of terms; or, multiply by the whole number of terms, and take half the product.

1. How many strokes does the hammer of a clock strike in 12 hours?

Half number of terms $=6$.
Extremes $1+12=13$. $13\times6=78$. Ans.

PROBLEM II.—The first term, the number of terms, and the common difference being given to find the last term.

RULE.—Multiply the common difference by the number of terms less 1, and add the product to the first term; or, subtract it from it, according as the series is ascending or descending.

2. The first term of a series being 3, the common difference 2, and the number of terms 7; what is the last term?

Number of terms 7; less $1=6$; $6\times2=12+3=15$, the last term.

3. The first term of a descending series being 15, the number of terms 7, and the common difference 2, required the last term.

$2\times6=12$. $15-12=3$, the required term. Ans.

4. The rent of a house being $50 a year, remained unpaid for 3 years; what was then due, simple interest being allowed at 6 per cent.?   Ans. $558.

5. A triangular brick pavement is to be formed, so that, commencing with 1 brick, the rows shall increase by 2 bricks, and shall be 88 in all. How many bricks in the last

row, how many in all, and if disposed in the form of a square, how many bricks will there be on a side?

$$\text{Ans.} \left\{ \begin{array}{l} \text{175 in the last row.} \\ \text{7744 in all.} \\ \text{88 bricks on a side.} \end{array} \right.$$

PROBLEM III.—The first term, the last term, and the number of terms given, to find the common difference.

RULE.—Divide the difference of the extremes by the number of terms less 1, and the quotient will be the common difference.

6. In a school there are 8 scholars, whose ages differ alike; the youngest is 4 years old, and the eldest 18; what is the common difference of their ages?

Ans. 2 years.

PROBLEM IV.—Given the first term, last term, and common difference, to find the number of terms.

RULE.—Divide the difference of the extremes by the common difference, and the quotient increased by 1 is the number of terms.

7. A man travelling a journey, went 18 miles the first day, and increased his distance each day by 2 miles; and the last day went 48 miles. How many days did he travel, and what distance?

$$\text{Ans.} \left\{ \begin{array}{l} \text{16 days.} \\ \text{528 miles.} \end{array} \right.$$

GEOMETRICAL PROGRESSION.

WHEN a series of numbers increases by a common multiplier, or decreases by a common divisor, it constitutes a *geometrical progression.*

Thus 2, 4, 8, 16, 32, increases by the common multiplier 2; and 81, 27, 9, 3, 1, decreases by the common divisor 3.* The common multiplier or divisor is called the

---

* A decreasing series may be produced as well by *multiplication* as division. Such a series does, in fact, always result from multiplication, when the ratio is less than unity; thus, $81 \times \frac{1}{3} = 27 \times \frac{1}{3} = 9 \times \frac{1}{3} = 3 \times \frac{1}{3} = 1$

ratio. The numbers of such a series are *continual proportionals,* there being the same ratio between every two consecutive terms. Consequently, every 4 consecutive terms constitute a proportion.

PROPOSITION.—The product of the extremes of a geometrical progression is equal to the product of any two equidistant terms; and likewise, to the square of the middle term. Thus, in the series 1, 2, 4, 8, 16, 32, 64; $\overline{2\times32}$, $\overline{4\times16}$, $(8)^2 = \overline{1\times64}$.

PROBLEM I.—Given the first term and the ratio, to find any proposed term.

RULE.—Raise the ratio to a power, whose index is 1 less than the number of the term sought; and multiply or divide the 1st term by that power, according as the series is ascending or descending.

1. If the first term be 2, and the ratio 3; what is the 5th term?

$5-1=4$, the number of the required term less 1.

The ratio $(3)^4 = 81\times2 = 162$. Ans.

The whole series is 2, 6, 18, 54, 162; which may be analyzed thus:

2, (2×3), $(2\times3\times3)$, $(2\times3\times3\times3)$, $(2\times3\times3\times3\times3)$:
That is, 2, $2\times(3)^1$, $2\times(3)^2$, $2\times(3)^3$, $2\times(3)^4$.

2. If the first term of a descending series be 162, and the ratio 3, required the 5th term.

$(3)^4 = 81$. $162 \div 81 = 2$ the required term.

PROBLEM II.—The extremes and ratio being given to find the sum of the series.

RULE.—Multiply the greater extreme by the ratio, from the product subtract the less extreme, and divide the remainder by the ratio less 1.

3. What is the sum of the series 2, 6, 18, 54, 162, 486, 1458?

Here, the ratio is 3, and the greater extreme 1458.

$\overline{1458\times3} - 2 = 4372 \div 2$ (the ratio less 1,) $= 2186$. Ans.
Why this produces the answer, will be made evident, if

we write down the series, and multiply it by the ratio, setting the product of each term under the next higher term, and then subtract the first series from the second.

2, 6, 18, 54, 162, 486, 1458.
6, 18, 54, 162, 486, 1458, 4374.

Remainder 4374—2 ; or 4372.

Here every term of the second series disappears by subtraction, *except the last*, which is the product of the greater extreme of the given series by the ratio.

Now, as we multiplied the given series by 3, and subtracted it from the product; it is obvious, that the remainder is equal to *twice the given series*. We therefore divide it by two.

Had the ratio been 4, the remainder would have been 3 times the given series ; that is, in all cases 1 time less than the number expressed by the ratio.

4. How many ancestors has every person reckoning twenty generations backward? and how many of the twentieth degree?

Ans. $\begin{cases} 2097150 \text{ in all.} \\ 1048576 \text{ of the 20th degree.} \end{cases}$

In this question there is given, the first term (2) ; the ratio (2), and the number of terms (20) to find the last term, and the sum of the series.

5. What debt can be discharged in a year, by paying 1 cent the first month, 3 cents the 2d, 9 the 3d, and so on in that ratio, for 12 months ?   Ans. $ 2657.20.

6. What would be the produce of a kernel of wheat in 11 years, at 20 fold, the produce of each year being sowed the next ; allowing 5000 kernels to a quart ?

Ans. 64000000 bushels.

This answer is the produce of the *last year*. The whole may be found by finding the sum of the series by Problem II.

PROBLEM III.—The extremes and the number of terms being given to find the ratio.

RULE.—Divide the greater extreme by the less, an?

extract that root of the quotient denoted by the number of terms less 1.

7. The extremes of a geometrical series being 4 and 64, and the number of terms 5; what is the ratio?

$64 \div 4 = 16.$

Number of terms 5; less $1 = 4.$

The 4th root of $16 = 2$ the ratio.

PROBLEM LV.—To find a mean proportional between two given numbers.

RULE.—Extract the square root of their product.

8. What is the mean proportional between 4 and 16?

$$4 \times 16 = 64 : \sqrt{64} = 8. \text{ Ans.}$$

Here, 4, 8, 16 are a geometrical series, of which 8 is the mean between the extremes.

9. What is the mean proportional between 9 and 81?

Ans. 27.

ANNUITIES AT SIMPLE INTEREST.

AN annuity is a sum paid annually; or, at equal stated periods.

When the debtor keeps the annuity in his own hands beyond the time of payment, it is said to be in arrears.

The sum of all the annuities for the time they have been forborne, together with the interest due on each, is called the amount.

If an annuity is bought off, or paid all at once at the beginning of the first year, the price which is paid for it is called the present worth.

To find the amount of an annuity at simple interest.

RULE.—Find the interest of the given annuity for 1 year;* then for 2, 3, &c., years, up to the given time less 1 year. Multiply the annuity by the number of years given,

---

* Or, for half a year, if the annuity is to be paid semi-annually.

and add the product to the whole interest, and the sum will be the amount sought: or, multiply the interest on the annuity for one year, by the number of years less 1 ; add this product to the multiplicand, and multiply the sum by half the first multiplier : the product will be the whole interest. Then multiply the annuity by the number of years, and add the result to the last product.

1. If a store be rented for 5 years at $ 300 a year, and payment be forborne until the end of the period, what will be then due, the interest being at 7 per cent. ?

The rent for the first year is forborne 4 years after it falls due ; the rent for the second year is forborne 3 years ; for the third, 2 years ; for the fourth, 1 year ; and for the last year, is paid when it falls due. Interest, therefore, is cast *for the number of years less* 1.

First Method.

Interest for 1 year =$21
    "   " 2   " = 42
    "   " 3   " = 63
    "   " 4   " = 84 .
                ———

Whole interest - $ 210
$300 \times 5=$ - - 1500
                ———

Ans. $1700

Second Method.

Interest for 1 year =$21
No. of years $5-1 =$ 4
              ———

             84   Annuity - - 300
Add multiplicand - 21   No. of years -   5
           ———   ———

          105   1500
½ of 4=2 - -   2   Interest - 210
          ———   ———

Whole interest - $210   Ans. $ 1700

   It will be perceived that the sums, denoting the interest for the successive years, constitute an arithmetical progression; and that the second method is merely finding the sum of the series.

2. A house was leased for 7 years at $ 400 per annum, and the rent unpaid until the end of the lease; how much was then due, simple interest, at 6 per cent.?

Ans. $3304

3. What will be the amount of an annuity of $ 50, to be paid annually, but forborne 20 years; simple interest, at 6 per cent.? Ans. $ 1570.

ANNUITIES AT COMPOUND INTEREST.

I. The annual amounts of any sum at compound interest, constitute a geometrical series, of which the first term is the principal, and the ratio the amount of one dollar for a year.

4. If an annuity of $125 be forborne 4 years, what will be its amount at 6 per cent., compound interest?

RULE.—Raise the ratio to the power indicated by the number of years: multiply that power by the annuity, from that product subtract the annuity, and divide the remainder by the decimal part of the ratio.*

Ratio $(1.06)^4 = 1.2627 \times 125 = 157.8087$
                                    125
                                    ___

.06)32.8087(=$ 546.81 Ans.

5. What in the amount of an annuity of $260, forborne 3 years, at 7 per cent., compound interest?

Ratio $(1.07)^3$ Ans. $835.874

6. Find the amount of an annuity of $50, in arrears for 20 years, compound interest at 5 per cent.

Ans. $1653.29.

II. To find the present worth of an annuity at compound interest.

RULE.—Raise the ratio to the power denoted by the number of years, and divide the annuity by that power: subtract the quotient from the annuity and divide the remain- by the decimal part of the ratio.

---

* By reference to Problems I. and II. of Progression, it will be seen, that this rule is but the process of finding the last term and the sum of a geometrical series, when the first term, the ratio, and the number of terms are given.

.7. What ready money will purchase an annuity of $125, to continue 4 years at 6 per cent. ?

$(1.06)^4 = 1.2627)$ 125.0000(98.99
            113643

       113570
       101016
                         125.
                         98.99.
       125540
       113643
                       .06)26.01
       118970
       113643 $433.50. Ans.

8. What ready money will purchase an annuity of $ 100, to continue 3 years, at 7 per cent., compound interest?

                               Ans. $ 262.72

These two rules being of little practical use, examples under them are not multiplied. Those who have occasion for such calculations, will very much facilitate them by constructing tables, exhibiting the amounts, and the present worth of $ 1 or £1 payable annually at compound interest, for a series of years, at the required rate per cent. Multiplying the tabular amount or present worth of $1 for the given number of years by the annuity, will give the answer.

MENSURATION OF SURFACES.

ALL surfaces of whatever figure, are reduced or referred to squares for measurement. The annexed figures, one of which, is a Square, and the other a rectangle or Right-angled parallellogram, are made up of smaller divisions, all of which are squares.

fig. 1.   fig. 2.

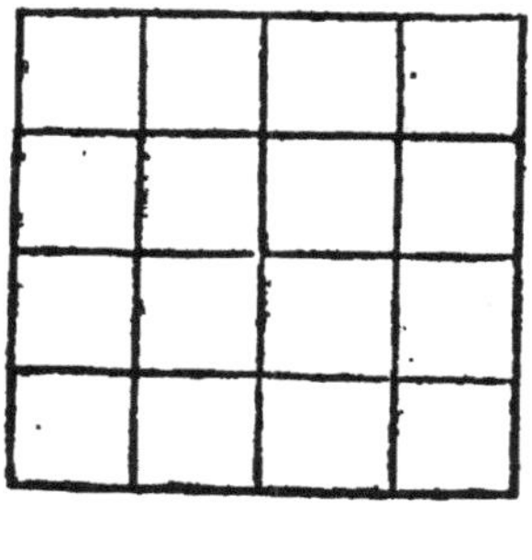

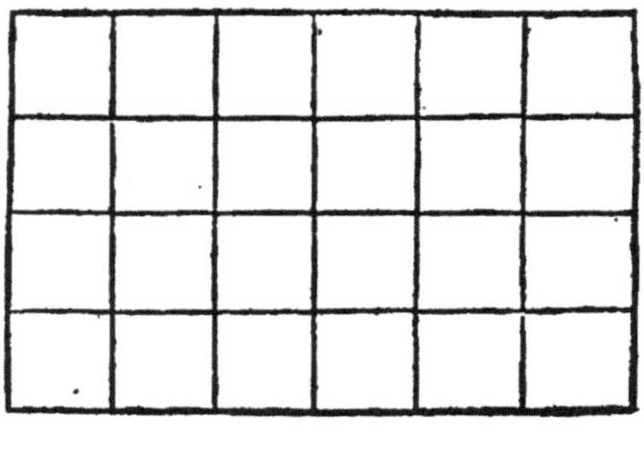

To measure or determine the area of such figures, is to find the number of equal parts of which they are made up ; the parts themselves being always *squares*.

Now it is obvious to inspection, that the whole number will be found, by multiplying two adjacent sides into each other. Thus in the rectangle, there are 6 divisions one way, and 4 the other : $6 \times 4 = 24$.

In a square, all the sides being equal, multiplying one side into another, is simply *squaring a side*.

For the purpose of measurement, the divisions may be large or small, provided they be exact squares, and equal to each other. Thus in the preceding figures, we may put 4 of these equal parts together, and they will still form a square. There will of course be fewer of them.

There are two other figures, which have a direct relation to these, in the comparative *lengths* of their sides, but differ in the *position* of them. These are the *Rhombus* and the *Rhomboid.*

The rhombus, like the square, has all its sides equal, but its angles are not *right-angles.*

Fig. 3.

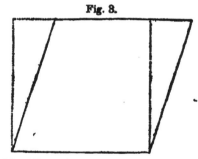

The rhomboid, like the rectangle, (or oblong,) has its opposite sides equal and parallel, but its angles are not right-angles ; in consequence of which, the adjacent side are not perpendicular, but oblique to each other.

Fig. 4.

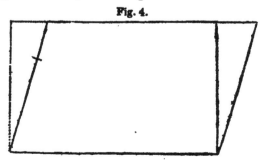

These two figures may be conceived to be formed from the others, by supposing their bases to be fixed, and the other sides to be moved on the lower angles into an oblique position.   This, it is evident, would bring the base and opposite side nearer to each other, just in proportion to the degree of obliquity.   It follows, therefore, that the space enclosed by the lines in that position, is less than when they were all perpendicular.

A rhombus, therefore, is less than a square of equal sides, and a rhomboid than a rectangle.

Their area is only equal to that contained by a rectangle of a breadth, equal to the perpendicular distance from the base of the rhombus or rhomboid, to its opposite side, and of a length equal to the base of those figures respectively. The rectangle which measures each of these figures, is indicated in part, by the dotted lines ; one of which encloses an area, not within the figure, just equal to the portion of it cut off by the other.

Therefore, to find the area of a rhombus or rhomboid :

RULE.—Multiply the length by the perpendicular breadth.

A TRIANGLE is a figure of three sides and three angles.

If two of the sides are perpendicular to each other it is called a right-angled triangle. Thus, A, B, C, has a right-angle at B. By drawing lines parallel to the two sides adjacent to the right-angle, (as the dotted lines,) a rectangle is formed, and the hypotenuse bisects* it. The area of the triangle is therefore, half that of a rectangle of equal sides.   Consequently,

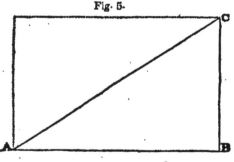

Fig. 5.

To find the area of a right-angled triangle :

RULE.—Multiply the base by the perpendicular, and take half the product ; or, multiply one into half the other, and the result is the area.

If the triangle be not right-angled, that is, have not two

---

* Cuts it into two equal parts

of its sides perpendicular to each other; as Fig. 6, we may determine the area, if we can get the perpendicular distance from the vertex to the base (C. D.)

RULE.—Multiply half that distance into the base.

Fig. 6.

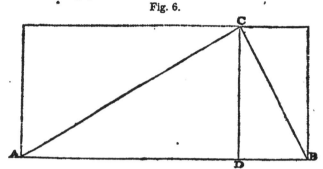

The completion of the rectangle (shown by the dotted lines) makes the reason of the rule apparent.

In case the perpendicular distance from the base to the opposite angle, cannot be obtained, the following method may be adopted:

RULE.—Add the three sides together, and take half their sum; from that half sum subtract each of the sides separately; multiply the three remainders thus obtained, and the half sum, continually into each other, and the square root of the last product, will be the area of the triangle.

NOTE.—The demonstration of this rule, would not be intelligible without a knowledge of geometry.

Fig. 7.

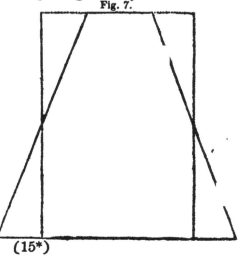

When the vertex of a triangle is cut off by a line parallel to the base, the figure thus formed is called a trapezoid. Thus, Fig. 7, is a trapezoid. The rectangle indicated by the dotted lines, is the measure of its area. It includes a surface without the figure, precisely equal to the portion excluded within the figure. The

(15*)

breadth of the rectangle is equal to half the sum of the two parallel sides.

RULE.—Multiply half the sum of the two parallel sides, by the perpendicular distance between them.

All the figures we have considered, have been referred either to a square or to a rectangle. The area of a circle is obtained in a similar manner. If we draw a square about a circle, so that each side shall touch its circumference, the diameter of the circle drawn from one of the points of contact, will meet the circle at the point of contact of the opposite side of the square.

Now, if we square the diameter of a circle, it is evident we shall have the area of a square, one side of which is equal to that diameter; or, in other words, of a square, which if drawn about the circle, (in the manner of the diagram,) would touch it in four points. But this would exceed the area of the circle, by the spaces included between its circumference, and each of the four angles of the square. It has, therefore, been an interesting problem with mathematicians, to ascertain the ratio between the area of a circle, and of a square drawn about it; and it is demonstrated, that the area of the circle is to that of a square, as the decimal .7854 is to unity.* If, therefore, the square contain 1 yard, a circle of that diameter would contain 7854 ten thousandths of a yard. If the square contain 4 yards, (its side being 2 yards,) then a circle of 2 yards in diameter, would contain .7854 × 4=3.1416 yards.

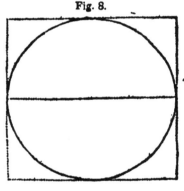

Fig. 8.

RULE.—Multiply the square of the diameter by the decimal .7854, and the product is the area.

When the circumference of a circle is known, the diameter may be found by the following proportion; as 22 : 7 : : so is the circumference to the diameter; or, as 3.1416 : 1 : :, &c.

---

* Nearly.

The converse of this statement will give the circumference, when the diameter is known.

DEFINITION.—A cylinder is a round body of uniform diameter, its two ends being parallel, and its axis perpendicular to them.

To find the area of a cylinder :

RULE.—Multiply the perimeter by the length.

To find the area of a sphere or globe :

RULE.—Multiply the circumference and diameter together; or, multiply the square of the circumference by 3.1416

QUESTIONS UNDER THE FOREGOING RULES.

1. What is the difference in quantity, between two pieces of land, one of which is a square with sides 40 rods in length, and the other a rhombus, with equal sides, but of a breadth from the base to the opposite side of only 34 rods ?
Ans. The square contains 240 rods the most.

2. There is a square floor, 18 feet on a side; what number of square feet does it contain; and how many yards of carpeting, 1 yard wide, will cover it ?
Ans. $\begin{cases} 324 \text{ feet.} \\ 36 \text{ yards.} \end{cases}$

3. There is a barn 50 feet by 36 : the body is 20 feet in height to the eaves. How many feet of boards will make the siding of the main body; and if the boards were all 15 inches in breadth, and 10 feet in length, what *number* of boards will it require ?
Ans. $\begin{cases} 3440 \text{ feet.} \\ 275. + \text{ boards.} \end{cases}$

4. On a base of 120 rods in length, a surveyor wished to lay off a rectangular lot of land, to contain 60 acres, what distance in rods must he run out from his base line ?
Ans. 80 rods.

5. How many square yards in a triangle, whose base is 48 feet, and perpendicular height $25\frac{1}{4}$ feet ?
Ans. $67\frac{1}{3}$ yards.

6. The three sides of a triangular field are severally 60, 50, and 40 rods; how many acres does it contain ?
Ans. 6 acres 32 rods.

7. How many acres in a triangular piece of land, whose sides are 30, 40, and 50 rods ?   Ans. 3¾ acres.

8. What is the difference between the area of a square, whose sides are 64 feet, and a rhombus of the same dimensions, whose perpendicular height is 52 feet ?
                                                            Ans. 768 feet.

9. A man bargaining for a piece of land of a rectangular shape, the length of which was 198 rods, and its breadth 150 rods, agreed to give $ 32 an acre, provided the owner would deduct from the quantity, the contents of a circular pond in it, which was admitted to be 100 rods in circumference. What was deducted for the pond, and what did the form come to?   Ans. $\begin{cases} 795 \text{ rods deducted.} \\ \$ 5781 \text{ the cost of the farm.} \end{cases}$

10. If the forward wheels of a coach are 4 feet, and the hind ones 5 feet in diameter; how many more times will the former revolve than the latter, in going a mile ?
                                                            Ans. 84 times.

11. If the diameter of a circular cistern be 6 feet; what is the area of its bottom ?   Ans. 28.2 feet.

12. What quantity of land in a field of the form of a trapezoid, the parallel sides of which are severally 60 rods and 40 rods, and their distance asunder 50 rods ?
                                                            Ans. 15⅝ acres.

13. How many square feet in a board 2 feet wide at the larger, and 1 foot 8 inches at the smaller end, and 14 feet long?   Ans. 25⅔ feet.

14. If you wish to plant 600 apple trees in a rectangular form, and are limited in one direction to 20 trees; how many trees must you have in the other direction to complete your orchard ?   Ans. 30 trees.

15. What is the quantity of surface in a solid cylinder, 20 feet long and 2 feet in diameter; the area of the two ends included ?   Ans. 131.9 feet.

NOTE.—The *perimeter* is the *measure round;* that is, the circumference. The area of the ends is to be calculated separately, and then added.

16. What is the convex area of a 12 inch globe; and suppose it to be of the uniform thickness of ¼ of an inch; what is the concave area of the same globe ?
                                      Ans. $\begin{cases} 3.1416 \text{ feet.} \\ 2.883 \text{ feet,} \end{cases}$

MENSURATION OF SOLIDS AND CAPACITIES.

As surfaces are measured by squares, solids are measured by cubes.

A cube is a solid of six equal and parallel sides, all of which are squares, and its angles all right-angles.

NOTE.—No school should be without a set of geometrical solids, like those in Holbrooks' apparatus. A bare inspection of these will give a better idea of the regular solids, than the most exact definition, or the most minute description.

The method of finding the solidity of a cube has already been given. See page 156.

1. How many solid feet in a square block 6 feet long, 6 broad, and 6 high? Ans. 216 feet.

2. There is a square cistern of uniform length, breadth, and depth, it being 8 feet in each of these dimensions. How many cubic feet does it contain; and allowing 231 inches to the gallon; what is its capacity in gallons?

Ans. $3330\frac{2}{77}$ gallons.

A PARALLELOPIPED is a solid of six quadrilateral sides, whereof every opposite two are parallel, equal, and similar.

To find its solidity:

RULE.—Multiply its length, breadth, and thickness continually together.

3. The plate supporting the rafters of a house, being 40 feet long, 14 inches wide, and 8 inches thick; how many solid feet does it contain? Ans. $31\frac{1}{9}$ feet.

4. If a pile of wood be 60 feet long, 10 feet high, and 3 feet wide; what number of feet and cords does it contain?

Ans. $\begin{cases} 1800 \text{ feet.} \\ 14\frac{1}{16} \text{ cords.} \end{cases}$

5. The bin of a granary is 10 feet long, 5 feet wide, and 4 feet high: allowing the cubical contents of a dry gallon to be $268\frac{4}{5}$ inches, how many bushels of grain will it contain? Ans. $160\frac{7}{10}$ bushels.

6. If you wanted a bin to contain twice as much, with a length of 12 feet, and a breadth of 6 feet; of what height must it be? Ans. $5\frac{5}{9}$ feet.

Note.—If you multiply the solid contents of the first bin by two, and divide by the product of the two given dimensions into each other, the quotient will be the answer.

A prism is a solid, whose ends are any similar, equal and parallel surfaces; and its sides rectangles.

To find the solidity of a prism, or a cylinder:

Rule.—Multiply the area of one end by the length.

7. The common prism used in decomposing the solar beam, is about 8 inches in length; and each of its ends is an equilateral triangle, whose three sides are severally about $1\frac{1}{4}$ inch. How many solid inches in the whole prism? Ans. 5.3 inches.

Note.—Find the area of the triangle by the 2d Rule on page 173.

8. How many solid feet in a circular pillar 40 feet long, and with a uniform diameter of 2 feet 5 inches? Ans. 196.35 feet.

9. There is a circular cistern of uniform diameter, whose depth is 8 feet, and diameter 5 feet; what is its capacity, allowing 282 inches to the gallon; and how much would its capacity be increased by adding 6 inches to its diameter?

Ans. $\begin{cases} 962\frac{1}{2} \text{ gallons its capacity.} \\ 202 \text{ gallons 1 pint its increase.} \end{cases}$

10. A man bought a grindstone, which was 48 inches in diameter, and 5 inches in thickness, for 10 dollars. When he had ground off 3 inches from its circumference, his neighbor proposed to purchase it from him, giving a proportional part of the first cost for the residue; deducting 4 inches every way from the centre, allowed to be the limit to which it could be used. What should he have paid?

Ans. $ 7.60.

To square a cylinder, that is, reduce it to a parallel-opiped, with sides of equal breadth:

Rule.—Multiply the square of the semi-diameter by 2, and the square root of the product will be the breadth of each of its sides.

11. If a round stick of timber be 18 inches in diameter; what will be its sides when hewn 4 square?

$9 \times 9 = 81 \times 2$. 162. $\sqrt{162} =$. 12.7 inches. Ans.

12. If a round log, 16 inches in diameter, be hewn 4 square, what will be the breadth of one of its sides?

Ans. 11.3 inches.

A PYRAMID is a solid which decreases uniformly from its base, until it comes to a point.

The base of a pyramid may be either a square, a triangle, or a circle. Hence it is named a *square* pyramid, a *triangular* pyramid, or a *cone*. The point in which the pyramid ends is called *its* vertex; a straight line from which, through the centre of the pyramid to its base, is called its axis.

To find the solidity of a pyramid:

RULE. — Multiply the area of the base by $\frac{1}{3}$ of its height.

When the smaller end of a pyramid is cut off parallel to the base, the residue is called the *Frustum of a pyramid.* When its base is a square, the solidity of the frustum is found by the following:

RULE.—To the product of the side of the base into the side of the top, add $\frac{1}{3}$ of the square of their difference, and multiply that sum by the length.

When the base is a circle:

RULE. — Add together the squares of each of its two diameters; and multiply $\frac{1}{3}$ of the sum by the decimal .7854, and that product by the length.

To find the solidity of a globe or sphere:

RULE.—Multiply the circumference and diameter together, and that product by $\frac{1}{6}$ of the diameter. Or,

Multiply the cube of the diameter by the decimal .5236.

NOTE. — If but one of these dimensions is given, the other must be found by the rule before laid down.

13. What number of solid feet in a piece of timber, whose ends are squares, each side of the greater end being 15 inches, and of the less end 6 inches, and the length or perpendicular altitude 24 feet? Ans. $19\frac{1}{4}$ feet.

14. How many cubic feet of marble will be required for

a monument; the pedestal of which is 2 feet square, and the pyramid* resting upon it 18 inches square at the base, and 10 inches at the top, and 5 feet high?   Ans. $14\frac{107}{108}$ feet.

15. Suppose a round stick of timber to be 2 feet in diameter at its largest end, 10 feet in length, and diminishing uniformly to the other end, which is 1 foot in diameter; how many solid feet does it contain?   Ans. 18.3 feet.

16. If a cask, which is two equal conic frustums, joined at their bases, have its bung diameter 28 inches, the head diameter 20 inches, and length 40 inches; how many wine gallons† does it contain?   Ans. 53.3 gallons.

17. What is the solidity of a pyramid, whose base is 4 feet square, and its perpendicular height 9 feet?
Ans. 48 feet.

18. If a conic pyramid be 27 feet high, and its base 7 feet in diameter; how many cubic feet does it contain?
Ans. $346\frac{1}{2}$ feet.

MISCELLANEOUS EXAMPLES.

1. Which is greatest, $1\frac{3}{4}$, or $\frac{33}{35}$?

2. Which is most, $\frac{2}{3}$ of a rod, or $\frac{3}{4}$ of 12 feet?

3. If the dividend be $\frac{3}{7}$, the quotient 320; what is the divisor?

4. The divisor being $\frac{1}{4}$, the quotient $\frac{8}{17}$; what is the dividend?

5. The divisor being 4520, the dividend $\frac{5}{9}$; what is the quotient?

6. The divisor to a number is .02, the quotient is 420; what is the number?

7. What is the difference between $\frac{2}{3}$ of £1 and $\frac{4}{5}$ of a shilling?

8. A board is 9 inches wide; how much in length will it require to make 12 square feet?   Ans. 16 feet.

---

* This is a frustum of a pyramid.
† A wine gallon is 231 inches.

9. A stick of timber is 9 inches wide and 7 inches thick; how much in length is required to make 21 solid feet? .

Ans. 48 feet.

10. A lot of land containing 60 acres, is bounded by four sides; two of its opposite ones are parallel, one of them 28, the other 32 rods in length; what is the perpendicular distance between them? Ans. 320 rods

11. A man found a purse of money; after taking out $\frac{1}{3}$ and $\frac{1}{5}$ of it, he found there were 70 dollars left; how many was there at first? Ans. $150.

12. One sixth of a post is in the mud under water, $\frac{1}{4}$ in the water, and 70 feet above the water; what is the whole length of the post? Ans. 120 feet.

13. A can do a piece of work in 5 days, B can do the same in 20 days; how long will they be about it if they both work together? Ans. 4 days.

14. A and B can do a piece of work in 8 days; A can do it alone in 12 days; in what time can B do it alone?

Ans. 24 days.

15. A merchant bought 63 gallons of rum for $28.35; how much water must be added to reduce the first cost to 35 cents per gallon? Ans. 18 gallons.

16. $1600 were put at interest at 6 per cent. per annum, until it amounted to $2000; what was the time?

Ans. 4 years 2 months.

17. $\frac{7}{8}$ of a certain number exceeds $\frac{6}{7}$ of itself by 10; what is the number? Ans. 560.

18. Divide a prize of $10200 among 60 men, 6 subaltern officers, 3 lieutenants, and a commander; giving to each subaltern double the share of a man, each lieutenant 3 times as much as the subaltern, and to the commander double that of a lieutenant; how much is each one's share?
Ans. { Captain $1200.
{ Each man $100.

19. A grocer bought a cask of molasses, containing 42 gallons; 6 gallons leaked out, and he sold the remainder at 37½ cents per gallon, gaining 20 per cent.; what was the original cost per gallon? Ans. .257.

20. If 12 men in 8 days, working 9 hours a day, can do a piece of work; how many men in 18 days can do the same, working 6 hours a day? Ans. 8 men.

21. A canal contractor engaged to excavate 80 rods of
(16)

earth in 15 days ; after 30 men had been employed 6 days, 20 rods of the canal were completed. Does he require fewer or more men to complete the residue according to contract, and how many ?

Ans. He requires 10 more men.

22. The top of a liberty pole being broken off by a blast of wind, struck the ground 10 feet distant from the foot of the pole ; what was the height of the pole, supposing the length of the broken piece to be 26 feet ?   Ans. 50 feet.

23. A ladder 26$\frac{2}{3}$ feet long, may be so planted that it shall reach a window 22 feet from the ground on one side of a street, and by turning it over, without moving the foot, it will reach another window 14 feet high on the other side ; what is the whole breadth of the street ?

Ans. 37 feet 9$\frac{1}{2}$ inches.

24. An iron ball, 4 inches in diameter, weighs 9 pounds, and the weights being as the cubes of the diameters ; required the weight of a spherical shell of 9 inches in diameter, and 1 inch thick.   Ans. 54 pounds 4 ounces.

25. Falling bodies pass over 16 feet the first second of time, and 64 feet in two seconds, the space increasing as the square of the times ; how long will it require a body to fall 400 feet, and how long to pass over the last 76 feet ?

Ans. $\begin{cases} \text{5 seconds to fall 400 feet.} \\ \text{$\frac{1}{2}$ second to pass over the last 76 feet.} \end{cases}$

26. Mix corn at 56 cents per bushel, and rye at 60, with 4 bushels of oats at 30 cents per bushel, and 2 bushels at 36 cents per bushel, and make the whole mixture worth 58 cents a bushel. How much corn and rye must the mixture contain ?   Ans. $\begin{cases} \text{6 bushels of corn.} \\ \text{84 of rye.} \end{cases}$

NOTE.—First find the average price, or the price per bushel of 4 bushels of oats at 30, and of 2; at 36. Then make a mixture of 6 bushels of oats at that price, with corn and rye as required.

27. A man spent all his income and $\frac{1}{4}$ more the first year after he became of age ; afterwards, for 4 years, he saved each year a sum equal to $\frac{1}{11}$ of his income ; and then paying his debts he had 90 dollars left. Required his annual income.   Ans. $550.

28. The greatest term in an arithmetical series is 70 ;

the common difference 3, and the number of terms 21 : what is the least term and the sum of the series ?

Ans. $\begin{cases} \text{The least term is 10.} \\ \text{Sum of the terms 840.} \end{cases}$

29. A man owes $ 800, due in 4 years hence, without interest; what ought he to pay now, supposing money to be worth 5 per cent. ? Ans. $ 666.66$\frac{2}{3}$.

30. A man received $ 134.40 interest for the use of $ 420, the rate being 6 per cent.; how long a time was the money at interest ? Ans. 5 years 4 months.

31. The hour and minute hands of a watch are together at 12 o'clock; when are they next together ?

Ans. 1 h., 5 m., 27$\frac{3}{11}$ s.

32. The mean revolution of the moon round the celestial sphere is performed in 27 days, 7 hours, 43 minutes, and 3 seconds; and that of the sun in 365 days, 5 hours, 48 minutes, and 54 seconds; what time must elapse between two successive conjunctions or oppositions ?

Ans. 29 days, 12 h., 44 m., 3 s.

NOTE.—This problem is exactly similar to the one immediately preceding.

33. The mean revolution of the moon being as stated in the last problem; how much will she change her longitude in one day, moving at a mean rate ? Ans. 13° 10' 35''.

34. A hare is 50 leaps before a greyhound, and takes 4 leaps to the greyhound's 3; but two leaps of the hound are equal to 3 of the hare's; how many leaps must the greyhound take before he catches the hare ? Ans. 300.

Observe, the hare takes 8 leaps while the hound takes 6, but 2 of his leaps are equal to 3 of hers; therefore 6 of his equal 9 of hers.

35. If 1000 men, besieged in a town with provisions for 5 weeks, allowing each man 16 ounces a day, be reinforced with 500 men more; and supposing that they cannot be relieved until the end of 8 weeks; how many ounces a day must each man have, that the provisions may last that time ? Ans. 6$\frac{2}{3}$ ounces.

36. If 20 men can perform a piece of work in 12 days; how many men will accomplish another thrice as large, in a fifth part of the time ? Ans. 300 men.

37. Two persons, A and B, travel between New York and Philadelphia, starting at the same instant: after 7

hours, they meet on the road, when it appears that A had rode $1\frac{1}{2}$ mile per hour more than B. At what rate per hour did each travel, supposing the distance to be 100 miles? Ans. $\begin{cases} \text{A } 7\frac{21}{28} \text{ miles.} \\ \text{B } 6\frac{11}{28} \text{ miles.} \end{cases}$

38. Divide 1200 acres of land among 3 persons, A, B, and C, so that B may have 100 acres more than A, and C 64 more than B. Ans. $\begin{cases} \text{A } 312 \text{ acres.} \\ \text{B } 412 \text{ acres} \\ \text{C } 476 \text{ acres.} \end{cases}$

NOTE.—After B and C take their extra acres, they must share equally.

39. A father taking his four sons to school, divided 153 shillings among them. Now the third had 9 shillings more than the youngest; the second 12 shillings more than the third; and the eldest 18 shillings more than the second: How much had each?

Ans. 21, 30, 42, and 60 shillings respectively.

40. A cistern is filled in 20 minutes by three pipes, one of which conveys 10 gallons more, and the other 5 gallons less than the third *per minute*. The cistern holds 820 gallons. How much flows through each pipe in a minute?

Ans. 22, 7, and 12 gallons.

41. A person employed 4 workmen; to the first of whom he gave 2 shillings more than to the second; to the second 3 shilling more than to the third; and to the third 4 shillings more than to the fourth: their wages amounted to 32 shillings; what did each receive?

Ans. 12, 10, 7, and 3 shillings.

42. Suppose that I have $\frac{3}{16}$ of a ship worth £1200; what part of her have I left after selling $\frac{2}{5}$ of $\frac{4}{5}$ of my share, and what is it worth? Ans. $\frac{37}{240}$ worth £185.

43. A father divided his fortune among his 3 sons, giving A £4 as often as B 3; and C 5 as often as B 6: supposing A's share was £4000, what was the whole legacy?

Ans. 9500 pounds.

44. A person bought 180 oranges at 2 a penny, and 180 more at 3 a penny; he sold them out again at 5 for 2 pence; did he gain or lose by the bargain?

Ans. He lost 6 pence.

45. If a quantity of provisions serves 1500 men 12 weeks, at the rate of 20 ounces a day for each man; how many

men will the same provisions maintain 20 weeks, at the rate of 8 ounces a day for each man ?   Ans. 2250 men.

46. A father left his son a fortune, $\frac{1}{4}$ of which he run through in 8 months, $\frac{3}{7}$ of the remainder lasted him 12 months longer ; after which he had barely £ 820 left. •What sum did the father bequeath to his son ?   Ans. £ 1913 6s. 8d.

47. A younger brother received £ 8400, which was just $\frac{3}{7}$ of his elder brother's portion ; what was the father worth at his death ?   Ans. £ 19200.

48. A man being asked the time of day, said it was between 5 and 6 ; but a more particular answer being required, said that the hour and minute hands were exactly together. What was the time ?   Ans. $27\frac{3}{11}$ minutes past 5.

49. 100 eggs being placed on the ground in a straight line, 1 yard from each other ; how far will a person travel who shall bring them one by one to a basket, which is placed at 1 yard from the first egg ?

Ans. 10100 yards, or 5 miles and 1300 yards.

50. Divide £ 1000 among A, B, and C, so as to give A 120 more, and B 95 less than C.

Ans. $\begin{cases} \text{A 445.} \\ \text{B 230.} \\ \text{C 325.} \end{cases}$

51. A wall was to be built 700 yards long in 29 days ; 12 men employed on it for 11 days, completed only 220 yards of it. How many must be added to the former, that the whole number may finish the wall in the given time, at the same rate of working ?   Ans. 4 men to be added.

52. Two persons, standing on opposite sides of a wood, which is 536 yards about, start to go round it, both in the same direction ; A at the rate 11 yards per minute ; and B 34 yards in 3 minutes ; how many times will the wood be gone round, before the quicker overtake the slower ?

Ans. 17 times.

53. A can do a piece of work alone in 12 days, and B in 14 ; in what time can both together perform a like quantity of work ?.   Ans. $6\frac{6}{13}$ days.

54. A person, after spending £ 20 more than $\frac{1}{4}$ of his income, had then remaining £ 30 more than the half of it ? what was his income ?   Ans. £ 200.

BOOK-KEEPING.

On account of the contracted size of the page, it has been found necessary to omit the forms of the journal and the leger, used by tradesmen, in recording their business transactions; as there is not sufficient space for the entries, together with the necessary columns for money, &c.

A method, however, of keeping accounts is given, in which only one book is used, which will be found well calculated for the purposes of farmers, mechanics, and others, whose business is not extensive, and is confined mainly to one set of customers.

Observe, that the two right hand columns are the Cr., and those on the left hand of them, the Dr. columns.

Either at the beginning or end of the book, or in a small separate book, an *alphabet*, containing the names of all the individuals with whom an account is opened, together with the *folio*, where that account is recorded, will be convenient for the purpose of reference.

The form given on the right hand page will show the manner of making entries. The method of balancing an account will be seen on the next page.

It sometimes happens, that errors are committed in making entries; that, one is made Dr. to an article when he should have been Cr. In such cases the error should not be rectified by an *erasure*, which defaces the appearance of the book; but, by crediting one party and debiting the other the amount of the error.

Care should be taken to set the figures in the money columns, properly under one another; which will facilitate the additions and prevent mistakes.

PROMISSORY NOTES.

For value received, I promise to pay Cyrus Vintner sixty dollars seventy-five cents, three months after date, with interest.

Albany, March 4, 1830. JAMES STOCKTON.

Utica, April 18, 1832.

Six months after date, I promise, for value received, to pay Abraham Stewart, or bearer, thirty-five dollars, with interest. GEORGE BANKHEAD.

LANSING WATTS.

1836.				Dr.		Cr.	
				Dol.	Cts.	Dol.	Cts.
Jan.	15	To 1 pr. fine Boots - - - - - - - -		7	00		
"	18	" 2 " Ladies' Slips (per daughter) $1.25 - - - -		2	50		
"	20	By 5 cords Maple Wood, at $1.25 - - - - 6.25				7	75
"	"	" 5 bu. Potatoes - - 30 cts. - - - - 1.50					
"	31	To 1 pr. Boys' Shoes - - - - - - -		1			
Feb.	21	By 6 bu. Turnips - - 75 cts. - - - -				4	50
"	30	To 1 pr. Gaiter Boots - - - - - -		2			
Mar.	30	" mending Boots for self - - - - .50					
"	31	" " " son .25					

JOHN POMEROY.

1836.					Dr. Dol.	Cts.	Cr. Dol.	Cts.
Jan.	12	To 2 Cherry Bedsteads - - -	$4.50	$9.00	11	00		
"	"	" 1 Wash-stand - - -		2.00				
Feb.	3	" 1 Cherry Table - - -		5.00	8	00		
"	"	" 6 Chairs (common) - -	50 cts.	3.00				
"	"	By his order on John Holmes					10	00
"	25	" Repairing clock					2	00
Mar.	1	To 1 Rocking Chair - -		2.25	5	25		
"	"	" 1 Dressing Table - -		3.00				
"	31	By 1 Silver Watch - - -					11	00
April	15	To 1 Settee - - - -			5	00		
"	20	By Balance of acct. carried down					6	25
					29	25	29	25
April	20	To Balance brought down -			6	25		

SAMUEL JOHNSON.

1836			Dr.			Cr.	
			Dol.	Cts.		Dol.	Cts.
Mar. 4	To 5 loads of Wood - - - $1.25 - - $6 25						
" "	" 6 bushels of Wheat - - 1.00 - 6.00						
" "	" 3 " Turnips - - 18¾ .56		12	81			
" "	By 4 days' Work - - - .75					3	00
July 6	" 5 days' Mowing - 1.00					5	00
July 10	" 4 days' Harvesting - 1.00					4	00
Sept. 3	To 8 bushels of Corn - .50 - 4.00						
" "	" Horse and Wagon to go to mill - .50 .50		4	50			
Nov. 8	By 1 Cow					12	00
Dec. 1	To 12 bushels of Potatoes - 20 cts.		3	60			
" "	To use of Team to draw wood		1	00			
Dec. 1	To Balance of acct. -		2	09			
			24	00		24	00
Dec. 1	By Balance of old acct. -					2	09

INDEX.

CPSIA information can be obtained
at www.ICGtesting.com
Printed in the USA
BVHW060408061118
532208BV00021B/3409/P

9 780266 875901